Studien zur Mustererkennung

Band 49

Studien zur Mustererkennung

Band 49

Herausgegeben von

Prof. Dr.-Ing. Heinrich Niemann
Prof. Dr.-Ing. Elmar Nöth

Katharina Hahn

Statistical iterative reconstruction and dose reduction in multi-slice computed tomography

Logos Verlag Berlin

λογος

Bibliografische Information der Deutschen Nationalbibliothek

Die Deutsche Nationalbibliothek verzeichnet diese Publikation in der
Deutschen Nationalbibliografie; detaillierte bibliografische Daten sind
im Internet über http://dnb.d-nb.de abrufbar.

978-3-8325-5443-9
ISSN 1617-0695

Logos Verlag Berlin GmbH
Georg-Knorr-Str. 4, Geb. 10, 12681 Berlin

Tel.:   +49 (0)30 / 42 85 10 90
Fax: +49 (0)30 / 42 85 10 92

https://www.logos-verlag.de

Abstract

Currently, computed tomography (CT) is one of the most important imaging methods in medical technology. Although CT examinations only make up a small proportion of X-ray examinations, they do make a great contribution to the civilization radiation exposure of the population. In particular, the use of modern CT techniques has made it possible to lower the mean radiation dose per examination over the past few years. This also includes the use of statistical iterative reconstruction methods (SIRM). While SIRM enable the modeling of physical imaging properties, the user can also decide freely and independently about the choice of numerous free parameters. However, every decision regarding parameterization has an influence on the final image quality (IQ). Accordingly, the main goal of this work was to examine the influence of some SIRM parameters in a more detailed manner.

In this work, initially all essential and necessary mathematical relationships of the data acquisition in addition to the basics of SIRM in CT were explained. A description of the basics of the IQ assessment followed, including a quantitative evaluation using basic metrics and task-based metrics.

When defining a SIRM, the definition of the modeling of the forward projection model plays a central role. In a preliminary study, various linear interpolation methods and basis function techniques were examined. Based on these results, an extensive study based on linear interpolation techniques was carried out. The choice of the interpolation method fundamentally influences IQ. Depending on both the pixel size of the reconstructed volume and on the choice of the linear interpolation method, the reconstruction error can be reduced by more than 65%. However, the task-based IQ assessment showed that the differences observed were not statistically significant. Furthermore, the influence of statistical weighting matrices (SWM) together with two different regularization methods (RM) was examined. In the first approach, the SWM carried the information about differently shaped bowtie filters and, in the second approach, the SWM carried the information about the handling of redundant data. Regularization was carried out by specifying a maximum number of iterations and by using a stopping criterion based on the absolute change in pixel values in the reconstruction volume. Both factors, the definition of the SWM and the RM, have a considerable, sometimes unexpected, impact on the final IQ in terms of reconstruction error and noise effects, i.e., for a non-regularized SWM solution the mean standard deviation was reduced by 45% whereas that value increased by 11% for the regularized SWM solution. In summary, the investigation of some SIRM parameters shows that the definition of iterative reconstruction parameters is not always trivial and must always be understood comprehensively in order to obtain an optimal IQ.

Many currently available SIRM have relatively long reconstruction times due to their underlying design. For this reason, a SIRM called FINESSE was developed that combines the properties of the FISTA algorithm with the block-based descent method. FINESSE showed promising results in both a simulation study and in a study with real data.

Kurzfassung

Heutzutage zählt die Computertomographie (CT) zu einem der wichtigsten bildgebenden Verfahren in der Medizintechnik. Obwohl CT Untersuchungen insgesamt nur einen geringen Anteil der Röntgenuntersuchungen ausmachen, wird durch sie dennoch ein großer Beitrag zur zivilisatorischen Strahlenexposition der Bevölkerung geleistet. Insbesondere durch den Einsatz moderner CT-Techniken konnte die mittlere Strahlendosis pro Untersuchung über die letzten Jahre gesenkt werden. Hierzu zählt auch der Einsatz statistischer iterativer Rekonstruktionsverfahren (SIR). Einerseits ermöglichen SIR die Modellierung physikalischer bildgebender Eigenschaften, andererseits kann der Anwender selbstständig und unabhängig über die Wahl zahlreicher freier Parameter entscheiden. Jedoch hat jede Entscheidung der Parametrierung einen Einfluss auf die finale Bildqualität (BQ). Das Hauptziel dieser Arbeit war die genaue Untersuchung mehrerer Einflussfaktoren von SIR.

In dieser Arbeit wurden zunächst alle erforderlichen und notwendigen mathematischen Zusammenhänge der Datenakquise sowie die Grundlagen von SIR in der CT erläutert. Nachfolgend wird detailiert auf die Grundlagen der BQ-Beurteilung eingegangen, die eine quantitative Auswertung mittels grundlegender Metriken sowie aufgabenbasierter Metriken beinhaltet.

Bei der Festlegung eines SIR spielt die Definition der diskreten Modellierung der Forwärtsprojektion eine zentrale Rolle. In einer vorläufigen Studie wurden zunächst verschiedene lineare sowie auf Basisfunktionen basierende Interpolationstechniken (IT) untersucht. Daraufhin wurde eine umfangreiche Studie basierend auf rein linearen IT durchgeführt. Die Wahl der IT hat grundlegenden Einfluss auf die finale BQ. Eine Reduktion um mehr als 65% des Rekonstruktionsfehlers ist je nach Wahl der linearen IT möglich. Die aufgabenbasierte BQ-Beurteilung zeigte jedoch, dass die ermittelten Unterschiede statistisch nicht signifikant sind. Des Weiteren wurde der Einfluss statistischer Gewichtsmatrizen (SGM) im Zusammenhang mit zwei verschiedenen Regularisierungsmethoden (RM) untersucht. Die SGM trug im ersten Ansatz die Information über verschieden geformte Bowtie Filter und im zweiten Ansatz die Information über die Handhabung redundanter Messdaten. Regularisiert wurde einerseits durch die Vorgabe einer bestimmten Anzahl von Iterationen, andererseits mittels eines Stopkriteriums basierend auf der absoluten Veränderung der Pixelwerte im Rekonstruktionsvolumen. Sowohl die SGM als auch die RM haben erheblichen und teils unerwarteten Einfluss auf die finale BQ. Je nach Wahl von SGM und RM hat sich gezeigt, dass die mittlere Standardabweichung um bis zu 45% reduziert bzw. um bis zu 11% erhöht wird. Zusammenfassend zeigt die Untersuchung der Einflussfaktoren von SIR, dass die Festlegung von iterativen Rekonstruktionsparametern nicht immer trivial ist und stets intensiv verstanden sein muss um eine optimale BQ zu erhalten.

Viele aktuell verfügbare SIR haben relativ lange Rekonstruktionszeiten aufgrund ihres zugrundeliegenden Designs. Aus diesem Hindergrund wurde ein SIR, genannt FINESSE, entwickelt, das die Eigenschaften des FISTA Algorithmus mit der blockbasierten Abstiegsmethode kombiniert. FINESSE lieferte sowohl in einer Simulationsstudie als auch auf Anwendung realer Daten vielversprechende Ergebnisse.

Acknowledgment

The journey is the reward. - Confucius

The opportunity to work on a PhD in a cooperation project between the University of Erlangen-Nuremberg, Germany, the Utah Center of Advanced Imaging Research (UCAIR) in Salt Lake City, USA, and the physics group of the computed tomography department of Siemens Healthineers GmbH, Forchheim, Germany was a great opportunity that was accompanied by a lot of experience. My work would not have come to fruition without the help and support of my mentors, colleagues and friends. I would like to express wholehearted appreciation to anybody who has contributed to this thesis, lightened up my working days and made me enjoy the work. In the following, I would like to thank some people in particular:

First and foremost, Prof. Frédéric Noo, PhD, for giving me the opportunity to participate in this collaboration, for his invitation to work at UCAIR, for his loyalty, for his outstanding work as a mentor and advisor, for his never-ending enthusiastic encouragement, and for all the discussions that went on of years.

Prof. Dr.-Ing. Joachim Hornegger for supporting this special working opportunity and for giving me the opportunity to become a member of the medical image processing group.

The CT R&D CTC department of Siemens Healthineers GmbH, especially its department head Dr. Thomas Flohr, for initiating this project and for providing financial support for my research.

Dr.-Ing. Harald Schöndube for his constructive advice, sharing his experience, the provision of simulation and reconstruction software as well as the acquisition of measurement data from clinical CT devices and his help with a Siemens paper war.

Dr.-Ing. Andreas Maier for introducing me to my fellow PhD students on the Pattern Recognition Lab (LME) at the the University of Erlangen-Nürnberg and for making me feeling welcome.

I would also like to thank my fellow doctoral students, in particular Lukas Licko and Thomas Weidinger, as well as all other colleagues in the CT physics group at Siemens Healthcare in Forchheim for the always good working atmosphere and for their helpful comments and discussions. Special thanks to Zhicong Yu for helping me find my way around Salt Lake City, especially on my first trip.

And in the last place in this list, but in my heart my own family is right at the front. Many thanks to my dear husband Marius and our three little great children. I know you have occasionally had to go without things because of me over the past few years. So I am all the happier to have received so much support and love from you during this phase.

Hausen, January 3th 2022   Katharina Hahn

Contents

Chapter 8   FINESSE: a Fast Iterative Non-linear Exact Sub-space SEarch based Algorithm

CHAPTER 1

Introduction

Wilhelm Röntgen discovered X-rays on November 8, 1895. At that time, the research in that field was enormous, and is still. Less than a month later Röntgen's publication X-rays were used in medical imaging [Spie 95]. Since then, the number of X-ray examinations has increased steadily, i.e., five billion medical imaging examinations had been conducted worldwide by the year 2010 [Roob 10].

X-ray examinations are used in various fields in medical imaging, i.e., dental medicine, radiotherapy, and mammography. Computed tomography (CT) is just one field of the application of X-rays, but CT has become one of the most important tools for radiologists. However, even though X-ray examinations help radiologists in their diagnoses, each examination is accompanied with a certain radiation exposure to the patient that is said to be unhealthy. Therefore, a significant amount of research is still ongoing in the direction of reducing the patient's radiation dose.

In this chapter, Section 1.1 and 1.2 report the facts and concerns about CT examinations that motivated the topic of the present dissertation. A summary of the achieved scientific contributions to the progress of research follows. The chapter ends with an overview of the individual dissertation chapters in Section 1.4

1.1 Computed Tomography Examinations

The number of CT examinations has increased worldwide dramatically over the last two decades [Smit 09]. In the United States of America, approximately 72 million scans were performed in 2007 [De G 09]. In Germany in 2009, about 4.88 million people received at least one CT scan [hila 11]. The *Bundesamt für Strahlenschutz* reported an increase in CT examinations of about 40% between 2007 and 2016 [Bund 20]. The statistics of the Organization for Economic Co-operation and Development (OECD) provide the number of computed tomography scans per country in 2020 or latest available per 1 000 000 inhabitants for various countries [OECD 21]. According to their statistics, the number of CT scans is between 6 in Colombia and

Mexico and 111 in Japan (see Figure 1.1), whereas Germany sits somewhere in between with 35 CT scans.

The most likely reason for the popularity is that CT has many advantages over traditional 2D medical radiography. For example, interfering superimposition of structures outside the area of interest can be eliminated. In general, a three-dimensional volume is displayed using cross-sectional 2D images that are used for both diagnostic and treatment purposes. Depending on the diagnostic task, images can be viewed in axial, coronal, or sagittal planes and with different resolution levels. These advantages make CT an essential tool that allows reliable diagnoses for lesions, bone fractures, bleedings, bruises, swelling, and/or inflammation.

Nevertheless, each CT examination comes with an additional radiation exposure to the patient. Although only 9% of all X-ray examinations in Germany are done with CT, their overall contribution in total collective effective dose in 2016 was more than 65% [Bund 20]. The total radiation dose depends highly on multiple factors: volume scanned, number, and type of scan sequences, desired resolution, and image quality. The typical effective dose to the body for a chest X-ray scan is about 0.02 mSv, for a Head CT scan $1 - 2$ mSv, for an abdomen CT scan approximately 8 mSv, and for a cardiac CT angiogram $9 - 12$ mSv [Furl 10, Food 17]. Compared to the world average dose rate from naturally occurring sources of 2.4 mSv per year [Cutt 09], a single abdominal CT examination may easily add a radiation dose equal to three years of average background radiation.

1.2   Towards Low-Dose CT

Due to the fact that additional radiation dose may lead to damaged body cells, including DNA molecules [Bren 07], doctors always have to decide if a CT examination comes with a real benefit for the patient. Many publications state that there is an increased risk of cancer caused by the additional radiation dose in CT examinations. Some studies predict that, in the future, between three and five percent of all cancers worldwide will result from medical imaging [Food 17]. Another Australian study of 10.9 million people reported than one in every 1800 CT scans was followed by an excess cancer which in turn led to an increase of lifetime risk of developing cancer from 40.00% to 40.05% after a CT scan [Sasi 11, Math 13]. On the other hand, McCollough et al. [McCo 15] state in their publication that studies that indicate an increased risk of cancer are plagued with serious methodological limitations and several highly improbable results. They conclude that there is no evidence that low doses, i.e., radiation doses is below 2 mSv, can cause any long-term harm [Padd 14, Expe 14]. Baysson et al. also came to similar conclusions [Bays 12].

These controversial discussions have opened a wide field of research and development worldwide. Not only are charitable and medical organizations involved in special campaigns, but manufacturers of medical imaging devices also show real interest in attempting to lower patients' radiation dose with multiple approaches while keeping high image quality. Among many other examples, the development of detectors with little or no electronic noise, adapted bowtie filters, and the application of statistical iterative reconstruction may provide significant improvements toward obtaining that goal.

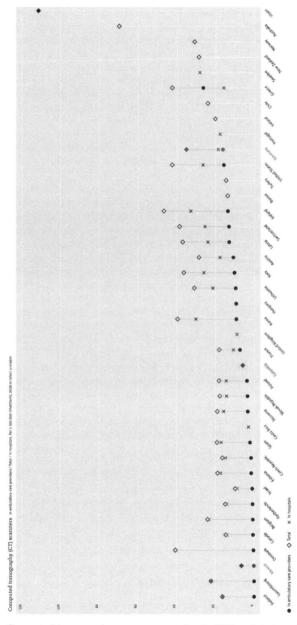

Figure 1.1: Computed tomography scans per country in 2020 or latest available per 1 000 000 inhabitants. This figure was published in [OECD 21].

Statistical iterative methods have long been commonplace, especially in nuclear medicine, since fewer data are available here and the image quality could be massively improved. In CT, analytical methods have long been the gold standard. But that has changed. The area of statistical iterative reconstruction has received a lot of attention in recent years. Unlike with analytical reconstruction methods, statistical iterative reconstruction methods have the advantage of allowing the modeling of any physical aspect of the CT data. In that way, they could be more exact than analytical methods. That advantage is coupled with the multiple degrees of freedom that iterative reconstruction methods have. Therefore, a thorough analysis of all possible influencing factors should be a natural part of each study. Up to now, to the best of the author's knowledge, there is a major lack of literature regarding fundamental investigations on the influence of single iterative reconstruction parameters on final image quality.

Another challenging factor in statistical iterative reconstruction is the reconstruction speed. Currently, almost all statistical iterative reconstruction methods are slow, making it difficult to use them in clinical routine. This is mainly due to the huge amounts of data that needs to be available during the whole reconstruction process. This can be easily understood with the following example based on the data sheet of the Siemens Healthineers CT scanner SOMATOM Drive. A typical chest or thorax examination includes up to 10 rotations. Depending on the scan settings, up to 2304 readings are transferred for the entire detector that is of size 736×32 pixels. Hence, the matrix size of the data is already of size $23040 \times 736 \times 32$ which corresponds to several Gigabytes depending which accuracy is preferred (float or double). In contrast to the analytical methods a so-called system matrix is required for all calculations. That matrix has the dimensions reading per rotation multiplied by the size of the detector times size of the reconstructed target volume. Most of the system matrix elements are zero, nevertheless the information about at least all non-zero elements need to be available during the iterative reconstruction process. Depending on the storage accuracy and design of the system matrix, several gigabytes of data are added. Overall, therefore, many gigabytes of memory may already be blocked. Having sufficient memory available is one problem, the other is that the system matrix must be available for every reconstruction. The calculation of the system matrix elements can either be done on the flow or the required information is read from the disk. Both options cost additional time. In addition, the reconstruction process is also slowed down by the fact that widely distributed elements of the data set have to be accessed again and again during the reconstruction.

Both, having a good understanding of the influencing factors in iterative reconstruction and handling the big data is essential to bring statistical iterative reconstruction in clinical routine. In the current dissertation, we try to fill at least some of these gaps.

1.3 Scientific Contributions

All evaluations were based on mostly MATLAB code, whereas all statistical iterative reconstructions were based on C code. For all iterative reconstruction algorithm implementations, no third-party software was used.

An overview of the original contributions of this dissertation along with the corresponding scientific publications is provided below. Please note that, due to my name change in 2014, there will be references to *Schmitt* (former name) and *Hahn*.

1. **Impact of Discrete Image Representation Techniques.** An essential basis in each statistical iterative reconstruction algorithm is the design of the forward projection model. Several forward projection models have been published over the last two decades. Which of the forward projection models the user should apply is still often a question of personal preference rather than based on quantitative evaluation using basic and/or task-based metrics. Chapter 6 tries to bridge that knowledge gap with an extensive investigation. For the quantitative evaluation, meaningful metrics had to be defined as part of Chapter 5. The results of that study were presented at four international conferences [Schm 12b, Schm 13, Schm 14a, Hahn 15a] and in one publication [Hahn 16].

2. **Challenges posed by Statistical Weights and Data Redundancies.** There are many ways to formulate a statistical reconstruction method for X-ray CT. In particular, the maximum likelihood solution without and with constraints on the image both appear highly popular. In the first approach, the user formulates an iterative algorithm that converges towards the maximum likelihood solution and defines the reconstruction as the application of a finite number of iteration steps. Using this approach, the iteration number is essentially seen as a regularization means. In the second approach, the regularization is not left to the iteration number; it is enforced directly by the constraint, and the user iterates as long as needed to reach the minimum of the objective function. The question which of these approaches is most appropriate in CT cannot be answered easily and needs to be analyzed thoroughly. In Chapter 7, we studied the effectiveness of regularization including essential aspects of CT imaging. The results of that investigation were presented at two international conferences [Schm 13, Hahn 15b].

3. **A Fast Iterative Non-linear Exact Sub-space SEarch based Algorithm (FINESSE).** Most existing statistical iterative reconstruction algorithms can be classified according to the number of voxels that are updated within one iteration. Some algorithms update only one voxel at a time, yielding an algorithm that converges quickly but is poorly parallelizable and thus time consuming. Other methods update all voxels simultaneously and are thereby much more amenable to parallelization, but these algorithms require many more iterations. Last, there are the third group of algorithms that aim at updating several but not all voxels within each iteration. These algorithms suffer from the drawback of requiring the solution of a complex sub-problem, which is often achieved using an approximate, monotonic update that slows down convergence. Currently, none of these algorithms is seen as being satisfactory for routine clinical usage. With the development of a novel algorithm, called *Fast Iterative Non-linear Exact Sub-space SEarch based Algorithm* (FINESSE), we want to fill that gap. First results have been shown on one international conference [Schm 14b]. The algorithm has also been patented [Schm 17].

1.4   Dissertation Outline

In the present dissertation, all background descriptions as well as comprehensive chapter-spanning contents were bundled in separate individual chapters. The main reasons for this were to facilitate readability and mainly to concentrate on the substantial content(s) of the subsequent chapters although the number of chapters was increased with that structure. Thus, Chapter 1 through Chapter 4 describe essential basics that will be referenced in subsequent chapters. Then, Chapter 5 to 9 will primarily focus on my own scientific contributions. A short description of the organization of the dissertation chapter-by-chapter is provided below.

Chapter 1 - Introduction

The introductory chapter starts with a description of the benefits of CT followed by the motivation of the dissertation. Finally, an overview of the original contributions as well as the organization of this dissertation is given.

Chapter 2 - Principles of Computed Tomography

The second chapter gives a brief introduction in the basic components of a computed tomography scanner. That is followed by a short summary of the development from the first clinically available CT scanner in 1972 up to modern commercially available CT scanners. Along with the evolution of the third generation of CT scanners, important gold standard techniques, such as scan modes, beam filtration, quarter detector offset, and flying focal spot were established. As most of those techniques serve as a basis of the conducted research and throughout the dissertation a rough description of them is given. That chapter concludes with the definition of the Hounsfield units.

Chapter 3 - Mathematical Description of the Data Acquisition Process

Chapter three covers the mathematical description of the measurement process that is used throughout the dissertation. Several coordinate systems, such as the world coordinate system, the detector coordinate system, and the image coordinate system, are introduced. Since the definition of the source coordinate system changes with the projection sampling geometry, this coordinate system is introduced separately for each geometry. The underlying coordinate systems are used to give a mathematical description for the calculation of a line integral, which is either called the Radon transform for fan-beam or cone-beam geometries.

Chapter 4 - Iterative Reconstruction Techniques

The basics of iterative reconstruction techniques, the description of the definition of the objective function, and the definition of a non-constrained and constrained image reconstruction problem are described in Chapter Four. Moreover, two commonly known iterative reconstruction methods, the Landweber method and the ICD method, are explained. Finally, the strengths and weaknesses of iterative reconstruction algorithms in general are discussed.

Chapter 5 - Image Quality Assessment

Image quality assessment is a central concept for evaluating the key performance parameters of a CT scanner, for example, resolution or noise. It is very important that the final reconstructed image is related to how well it conveys all anatomical or functional information. The signs of a disease or an injury need to be clearly visible so the interpreting radiologist can make an accurate diagnosis. Different quantitative metrics exist to evaluate certain IQ properties, such as basic metrics and task-based metrics, which are both described in detail in that chapter. Furthermore, details about the phantom used to evaluate image quality are given. This chapter ends with a discussion explaining the pros and cons for the basic and task-based IQ assessment for the choices being made throughout the dissertation.

Chapter 6 - Impact of Discrete Image Representation Techniques on Image Quality

In literature, many definitions of discrete image representation techniques that may affect the image quality of the final reconstructed images exist. Conceptually, the choice of the forward projection model is a major step in the design of an IR algorithm, particularly because the decision being made at this level affects both bias and noise properties of the reconstruction. In addition, the selection of additional parameters that might appear in the cost function or the regularization also influences image quality. However, optimizing a variety of parameters is complicated and accompanied with the need to perform several reconstructions to account for different noise realizations and variations in geometry, which is essential for meaningful observations. Therefore, the focus in Chapter 6 is exclusively on the impact of discrete image representation techniques.

This chapter starts with a mathematical formulation of various image representation techniques that have been extensively investigated. In order to allow a fair comparison of all methods, a description of the experimental comparison conditions is given. Based on the results of a preliminary study, three linear forward projection models, namely the Joseph's method, the distance-driven method, and the bilinear methods were selected to continue with a more detailed analysis including both basic and task-based metrics. The chapter concludes with a discussion and conclusion.

Chapter 7 - Challenges posed by Statistical Weights and Data Redundancies

Iterative reconstruction methods allow modeling of the properties of the line integral measurement using a statistical weighting matrix. In the current chapter, the statistical weighting matrix was formed by: i) taking the effect of different bowtie filters; and, ii) considering data redundancies. The solution of the iterative reconstruction method was then found using two different regularized reconstruction methods, namely the

Landweber method and the ICD method. The Landweber method was regularized by stopping after a certain number of iterates, whereas the ICD method was stopped when the change in pixel value of the reconstructed volume became smaller than a predefined number. Subsequently, the impact of the regularization method together with the respective statistical weighting matrix was investigated.

After a description of the experimental setup, the concept of the ICD parameter selection is given. The results are given in two subsequent sections where each of them describes the definition of the statistical weighting matrix, the parameter selection, and the results individually. Finally, the chapter ends with an overall summary discussion.

Chapter 8 - FINESSE: a Fast Iterative Non-linear Exact Sub-space SEarch based Algorithm

In Chapter 8, a novel reconstruction algorithm, called FINESSE, is presented. That algorithm tries to fill a gap between iterative reconstruction algorithms that update all voxels within one iteration and between algorithms that update all voxels simultaneously. Whereas the first group of algorithms is poorly parallelizable and thus time consuming, the second group of algorithms is much more amenable to parallelization, but also generally requires many more iterations.

This chapter starts with a description of the algorithm design. Then, details about the experimental setup and the results of the simulation study are given. The subsequent section contains the results obtained with real data. A discussion and conclusion completes that chapter.

Chapter 9 - Summary and Outlook

The final chapter provides a summary of the conducted research and the scientific progress achieved by the work presented in this dissertation. The chapter concludes with a perspective on open research challenges.

CHAPTER 2

Principles of Computed Tomography

Computed tomography is an important tool in medical imaging for generating a non-overlapping and non-invasive image of a single slice through the scanned object. Besides a single, two-dimensional image slice, it is also possible to generate a three-dimensional sequence of images. This sequence consists of adjacent two-dimensional image slices. Such three-dimensional images facilitate the examination of the physical configuration and expansion of anatomical structures and of abnormal tissue alterations in the volume of the scanned object. Since the introduction of computed tomography and its first clinical application in 1972, research and development has propelled technological progress in a variety of ways.

This chapter starts with a description of the setup of a CT scanner, followed by a description of Lambert-Beer's Law and an overview of the development of CT scanners. Next, Section 2.4 describes the sampling geometries. Subsequently, some practical aspects are described such as scan modes, X-ray beam filtration, quarter detector offset, and flying focal spot. The chapter concludes with the definition of Hounsfield units.

2.1   Setup of a Computer Tomograph

A typical setup of a modern CT scanner is shown in Figure 2.1. The gantry of a CT scanner hides the power supply, most of the electronics necessary for controls and data pre-processing, as well as the data acquisitions system. The data acquisition system consists of an X-ray source and a detector. Both are mounted opposite to each

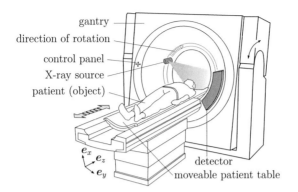

Figure 2.1: Typical setup of a modern CT scanner (3rd generation).

other on a rotating frame. The slip ring technique enables the transfer of electrical power and data so that cable entanglement through continuous rotation is avoided. This also allows taking measurements from any desired start angle.

In most CT scanners, the gantry can be tilted up to $\pm 30°$ with respect to the axis of rotation. This enables the taking of a CT measurement through structures of interest, i.e., the base of the skull or the lumbar spine. The gantry opening itself has a width of about 70 cm with a field of measurement of about 50 cm to allow also an examination of patients with a wide profile.

The patient lies on a patient table. The patient table is designed to be movable with a high level of positioning precision and in motion speed along the z-axis, i.e., orthogonal to the gantry rotation plane. During an examination, the X-ray source and the detector rotates with a constant speed with up to 4 rotations per second around the patient. The possibility of translating the patient table during the scan enables different scanning trajectories. A detailed description of possible scanning geometries is described in Chapter 2.4.

2.2   Lambert–Beer's Law

X-ray photons are absorbed or scattered when they pass through a material. This attenuation effect is due to interactions between the irradiated material and the photons, i.e., through the photoelectric effect, the Compton Effect, and the coherent scattering effect [Buzu 04, Hsie 09].

Let λ_x be the wavelength of photons of a monochromatic X-ray beam along the line $\mathcal{L}$, let $I_0(\lambda_x)$ and $I(\lambda_x)$ be the incident and transmitted X-ray intensity, and let $\mu(\boldsymbol{x}, \lambda_x)$ be the attenuation coefficient distribution of the irradiated material, where $\boldsymbol{x}$ denotes the position vector in Cartesian coordinates, i.e., $\boldsymbol{x} = [x, y]^T$ in two dimensions or $\boldsymbol{x} = [x, y, z]^T$ in three dimensions, respectively. The attenuation of photons that are traveling along the line $\mathcal{L}$ can be described by the Lambert–Beer's law that is given by

$$I(\lambda_x) = I_0(\lambda_x)\, e^{-\int_{\mathcal{L}} \mu(\boldsymbol{x}, \lambda_x)\, d\boldsymbol{x}} \;. \tag{2.1}$$

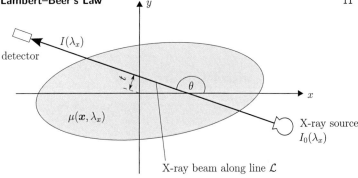

Figure 2.2: Mathematical model of Lambert–Beer's law for monochromatic emission. The effective intensity, I_0, from the x-ray source is attenuated by the material along $\mathcal{L}$. Finally, the remaining intensity, I, is measured by the detector. The parameters t and θ are used to describe the direction of the X-ray beam along the line $\mathcal{L}$ (see also Chapter 3).

An illustration of Lambert–Beer's law is shown in Figure 2.2. Note that Equation 2.1 represents a monochromatic measurement process whereas the output energy spectrum of the X-ray tube of a real measurement process is quite broad. To address this aspect, the measurements are pre-processed before any reconstruction is applied. The pre-processing step includes among others a correction for polychromatic X-ray radiation. After this correction step, the data can be assumed to be equivalent to a measured data set from a monochromatic X-ray radiation, meaning that all photons emitted by the source are of the same energy, called effective energy E_{eff}. Under this assumption Equation 2.1 simplifies to:

$$I = I_0\, e^{-\int_{\mathcal{L}} \mu(x)\, dx} \; , \tag{2.2}$$

where the attenuation coefficient μ is now adopted to be independent of the X-ray photon energy, i.e., μ varies only with x.

By applying the logarithm in Equation 2.2, a projection measurement, g, can be calculated. The measurement of g represents a single line integral over the attenuation coefficient along the X-ray path $\mathcal{L}$ and is given by

$$g = -\ln\left(\frac{I_0}{I}\right) = \int_{\mathcal{L}} \mu(\boldsymbol{x})\, d\boldsymbol{x} \; . \tag{2.3}$$

In CT, the attenuation coefficient distribution for an ensemble of lines is calculated based on solving Equation 2.3. Since the attenuation coefficient is heavily material dependent it is possible to distinguish materials from each other. Materials with a high μ value attenuate the X-ray beam more than materials with a low μ value, for example, the attenuation coefficient of bones is higher than that of soft tissue [Kale 06, Buzu 08, Kak 01].

Note that under the assumption of the ideal monochromatic Lambert–Beer's law different reconstruction artifacts may appear in the reconstructed images. For example, the beam hardening effect can cause cupping, shading, or streak artifacts. Note also that for all topics in this dissertation monochromatic radiation is assumed.

2.3   Generations of CT Scanners

This section provides information about the stages of development of different CT scanner generations. Historically, four generations of CT scanners have been developed over the last 50 years. In this section, only three of them are presented since the fourth generation of CT scanners is not in wide use. More information about the construction of a fourth generation scanner can be found in Kalender [Kale 06], Goldman [Gold 07], Buzug [Buzu 08], and Hsie [Hsie 09]. The demands of radiology have always driven fast progress from one generation to the next. This includes, for example, reduction of data acquisition time, reduction of X-ray exposure, improvement of image quality, reduction of costs, and optimization of the user interface.

2.3.1   First Generation CT Scanner

The first-generation CT scanner was built by Electric and Musical Industries Ltd. (EMI) in 1971 [Hsie 09]. The associated data acquisition system consisted of one detector element and an X-ray tube that emitted a single needle-like X-ray beam. The needle beam was created using an appropriated pinhole collimator. Within one view angle, both the X-ray source and the detector were translated perpendicular to the needle beam direction until the entire desired scanning field, also called field-of-view (FOV), was completely covered. Then, the whole data acquisition system was rotated by 1° and again translated. This process was repeated until data over 180 degrees for image reconstruction was acquired. Due to this translation-rotation system, only one pencil beam was measured at a time meaning that the scanning time was very long [Kale 06]. The principle of the first generation CT scanner is shown on the left in Figure 2.3.

The beam width in the original EMI head scanner was 3 mm within the plane of the slice and 13 mm wide perpendicular to the slice (across the scanning plane). Altogether, 160 measurements per view angle were taken. It required five to six minutes to complete a full scan of one single image slice [Gold 07].

2.3.2   Second Generation CT Scanner

CT scanners of the second generation were equipped with an X-ray source that emitted a narrow fan-beam. The single detector element was replaced by a short detector array consisting of about 30 elements [Buzu 08]. Due to the fan-beam, it was possible to measure multiple needle-like X-ray beams at the same time. Since the fan-beam angle was only about 10°, a translation-rotation system was still required to cover the total FOV. The right hand side in Figure 2.3 depicts the principle of a second generation CT scanner.

Due to the simultaneous measurement process of more than one X-ray beam at each translation location, it was possible to reduce the scanning time to approximately 20 s [Kale 06]. With this progress, it now became possible to perform body

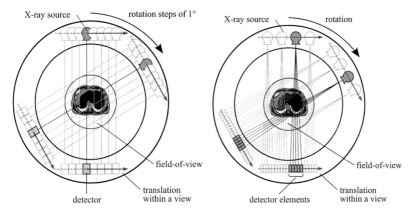

Figure 2.3: Principle of the first generation CT scanner. The data acquisition system was first translated linearly, and then the configuration was rotated by 1° *(left)*. A second generation CT scanner was equipped with a fan-beam X-ray source together with a short detector array. The translation-rotation system was still required *(right)*. Own drawing based on Buzug [Buzu 08].

scans within the breath-hold-range of most patients. In addition, the patient motion during data acquisition was drastically reduced and image quality was simultaneously improved.

2.3.3   Third Generation CT Scanner

In third generation CT scanners, the X-ray fan-beam angle is extended to $40-60°$ such that it becomes possible to simultaneously X-ray the entire FOV [Buzu 04]. Correspondingly, the detector array is also enlarged to intercept the X-ray beams at all times. Hence, the translation motion within a view is no longer necessary. In this new setup, the data acquisition system rotates synchronously around the patient and is also called the rotate/rotate geometry [Ulzh 09].

The arrangement of the detector arrays can be either that they form a flat detector or a curved detector as shown in Figure 2.4. In both cases, the line through the X-ray source and the rotation center is always perpendicular to the direction of the central detector element. In the case of the flat detector, all detector elements are aligned equidistantly along a straight line. The detector elements in a curved detector are arranged equiangular along a circle element. This circle element is concentric to the X-ray source position, i.e., the distance from the source to each point on the detector is the same. The angle between all lines connecting the X-ray source and the detector elements are of the same angular size. Due to their geometry features, the flat detector geometry is also called geometry of equidistant detector elements, whereas the curved detector geometry is also called equiangular ray geometry.

A typical third generation CT scanner has at least between 400 and 1000 detector arrays, a scanning FOV of usually 50 cm, and typically about 1000 measurements

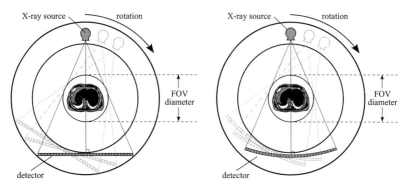

Figure 2.4: Third generation CT scanner geometry with a flat detector *(left)* and curved detector *(right)*. Both the fan-beam angle and the detector array are enlarged to cover the entire FOV. During a CT scan the data acquisition system rotates synchronously around the patient. Own drawing based on Buzug [Buzu 08].

over 360° are recorded [Gold 07, Ulzh 09]. The data acquisition time of today's third generation CT scanners is about one to four turns per second [Hsie 09]. Today, nearly all of the state-of-the-art scanners are third generation CT scanners because of the inherent advantages in terms of coverage of the FOV and sampling uniformity.

2.4   Projection Sampling Geometries

The projection sampling geometry is closely associated with the generation of the CT scanner. For the first and second CT generation, the rays of each projection angle are based on parallel measurements. This sampling method is called parallel-beam geometry or parallel-beam projection. The parallel-beam principle allows a straight forward mathematical description of the image reconstruction and is of fundamental importance for more than just this reason.

In the third CT generation, the data is collected using fan-beam geometry. In this geometry the samples of a single projection have the same focus, namely the X-ray source. The measurements are collected from all detector elements at the same time. Depending on the kind of the detector, the X-rays are either equidistantly or equiangularly sampled. An equidistant sampling is present when the detector is flat; an equiangular sampling is present when the detector is curved. For that reason, the fan-beam geometry may also be classified as an equidistant fan-beam geometry or an equiangular fan-beam geometry, respectively.

Besides the parallel-beam and fan-beam geometries, a third sampling geometry, the cone-beam geometry, is possible. In cone-beam geometry, multiple fan-beam planes are collected simultaneously to cover a volume [Hsie 09]. All fan-beam planes still have the same focus point, meaning that all planes beside the central fan-beam plane are tilted with respect to this plane. This sampling geometry is only possible in multi-row detectors.

a) parallel-beam   b) fan-beam   c) fan-beam   d) cone-beam
                (aquidistant)   (equiangular)

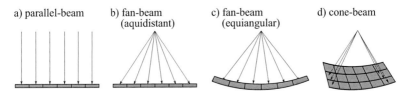

Figure 2.5: Projection sampling geometries.

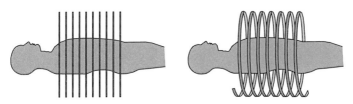

Figure 2.6: Scan modes of a modern CT scanner: conventional CT scan *(left)*, helical CT scan *(right)*.

The sampling geometries are depicted in Figure 2.5. The focus of this work is on parallel-beam and fan-beam geometry. The mathematical details of these geometries are discussed in Chapter 3.

2.5 Scan Modes

The principles of a conventional CT scan and a helical CT scan are illustrated in Figure 2.6. In the conventional scan mode, the data acquisition system rotates around the patient on a given z-location. After a full rotation, the table moves by a predefined distance. In most cases, this distance corresponds to the slice thickness. This procedure for a circular scan and the translation will be repeated until the desired distance in z is completely covered. In a spiral or helical scan mode, the patient will be moved continuously while the data acquisition system rotates. This means that the data acquisition system moves along a helix relative to the patient [Buzu 08].

A conventional CT scan mode can be carried out in a so-called short scan mode. Instead of a full rotation (full scan mode), only a partial of the full rotation is performed. Most short scan modes in conventional scan modes cover a rotation over 240°.

During a conventional CT scan, single slices are recorded. For that reason, it is not possible to perform an exact reconstruction in a plane outside of the scanned slice and this may lead to additional artifacts in the reconstructed image. However, for a single slice only one circular scan is needed. By performing a spiral scan, no gaps between the image slices appear due to the continuous shift of the patient table. Additionally, the scan time is usually shorter due to the steady movement of the table and the reduced sampling rate per single slice.

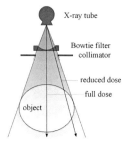

Figure 2.7: Schematic diagram of the X-ray filtration with a bowtie filter.

2.6   X-ray Beam Filtration

An X-ray tube emits a wide spectrum of photons with a wavelength below 10nm. This spectrum consists mostly of low energy photons. Since low energy photons are primarily absorbed by the patient, their contribution to the detected signal is very small. In order to reduce the patient's radiation dose near the edges of the scanning FOV, special X-ray filters are set directly behind the X-ray tube. Figure 2.7 shows a schematic diagram of the X-ray filtration using the so-called bowtie filter. A bowtie filter is a shaped piece of material (usually metal). It is designed to equalize the intensities of the rays hitting the detector for a given attenuating object. The effect of the bowtie filter is to make the number of photons going into the scanned object vary within the X-ray beam as indicated in Figure 2.7 [Hsie 09]. Another popular filter is the flat filter, which modifies the X-ray spectrum uniformly.

In general, X-ray filters are made of a material of low atomic number and of high density. Typical materials are for example aluminum and copper [Buzu 08].

2.7   Quarter Detector Offset and Flying Focal Spot

The finite size of the detector elements limits the resolution of the projections. To mitigate aliasing errors, the signal must be sampled at least twice within one detector element. More precisely, the samples must be measured in a distance smaller than or equal to half of the detector width. This is the well-known Nyquist sampling criterion [Hsie 09, Kak 01]. Both the quarter detector approach and the flying focal spot (FFS) approach allow an increase in the sampling rate such that the Nyquist sampling criterion is fulfilled for each of the methods and for the combination of the methods.

In the quarter detector approach, the detector cells are shifted by one-quarter detector element. Consequently, measurements that were taken over more than 180° result in the same line integrals, but measured on an interleaved grid. This concept is schematically shown in Figure 2.8. This approach effectively doubles the sampling rate for measurements over 360°.

a) quarter detector offset   b) flying focal spot

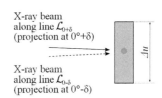

Figure 2.8: Schematic diagram for a CT scanner with quarter detector offset. The drawing shows the central detector element with a detector shift of one quarter. Opposite line integrals are measured in an interleaved grid *(left)*. Schematic diagram for a CT scanner with flying focal spot and with no quarter detector offset *(right)*.

In the flying focal spot approach, the focus of the X-ray tube is fast switched in angular direction by δ. This means that line integrals are generated between two closely neighboring points. Figure 2.8 displays the concept of the flying focal spot. As for the quarter detector offset approach, the sampling rate is again doubled.

Both concepts may be combined. Then, the detector offset needs to be an eighth of the detector element in respect to the iso-center. In that case, the number of samplings is quadrupled.

2.8   Hounsfield Units

In a reconstructed CT image, the attenuation values are converted into dimensionless values, $\mu_{\mathrm{HU}}(x)$, of the Hounsfield scale, HU [Hsie 09]. In this scale, the attenuation coefficients, $\mu(\boldsymbol{x}, \lambda)$, are normalized relative to the X-ray attenuation of water, $\lambda_{\mathrm{W}}(\lambda_x)$, at a given wavelength λ_x. The conversion formula is given by

$$\mu_{\mathrm{HU}}(x) := \frac{\mu(\boldsymbol{x}, \lambda_x) - \mu_{\mathrm{W}}(\lambda_x)}{\mu_{\mathrm{W}}(\lambda_x)} \cdot 1000\,\mathrm{HU} \ , \tag{2.4}$$

where μ_{HU} is also called CT value in dimensionless Hounsfield units. As a consequence of Equation 2.4, the attenuation of water is, for instance, 0 HU while the absence of attenuating material (air) corresponds to a CT value of -1000 HU. A more detailed classification of the Hounsfield units for different body structures is given in [Oppe 11]. Throughout the dissertation an attenuation value for water of $\mu_{\mathrm{W}} = 0.1836$ cm^{-1} is used.

a) quarter detector offset   b) flying focal spot

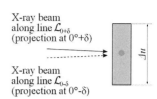

projection at 180°   projection at 0°

Figure 2.8: Schematic diagram for a CT scanner with quarter detector offset. The drawing shows the central detector element with a detector shift of one quarter. Opposite line integrals are measured in an interleaved grid *(left)*. Schematic diagram for a CT scanner with flying focal spot and with no quarter detector offset *(right)*.

In the flying focal spot approach, the focus of the X-ray tube is fast switched in angular direction by δ. This means that line integrals are generated between two closely neighboring points. Figure 2.8 displays the concept of the flying focal spot. As for the quarter detector offset approach, the sampling rate is again doubled.

Both concepts may be combined. Then, the detector offset needs to be an eighth of the detector element in respect to the iso-center. In that case, the number of samplings is quadrupled.

2.8   Hounsfield Units

In a reconstructed CT image, the attenuation values are converted into dimensionless values, $\mu_{\mathrm{HU}}(x)$, of the Hounsfield scale, HU [Hsie 09]. In this scale, the attenuation coefficients, $\mu(\boldsymbol{x}, \lambda)$, are normalized relative to the X-ray attenuation of water, $\lambda_{\mathrm{W}}(\lambda_x)$, at a given wavelength λ_x. The conversion formula is given by

$$\mu_{\mathrm{HU}}(x) := \frac{\mu(\boldsymbol{x}, \lambda_x) - \mu_{\mathrm{W}}(\lambda_x)}{\mu_{\mathrm{W}}(\lambda_x)} \cdot 1000\,\mathrm{HU}\ , \tag{2.4}$$

where μ_{HU} is also called CT value in dimensionless Hounsfield units. As a consequence of Equation 2.4, the attenuation of water is, for instance, 0 HU while the absence of attenuating material (air) corresponds to a CT value of -1000 HU. A more detailed classification of the Hounsfield units for different body structures is given in [Oppe 11]. Throughout the dissertation an attenuation value for water of $\mu_{\mathrm{W}} = 0.1836$ cm^{-1} is used.

CHAPTER 3

Mathematical Description of the Data Acquisition Process

The Austrian mathematician and former FAU professor Johann Radon first solved the mathematical problem of the two-dimensional image reconstruction from a line integral in 1917 [Rado 17]. Radon's results have been very important for modern CT because he gave a description of how the integral inversion can be efficiently done. The integral transform was named *Radon transform* in honor of Radon.

This chapter covers the mathematical description of the measurement process that is used throughout the dissertation. In Section 3.1, several coordinate systems, such as the world coordinate system, the detector coordinate system, and the image coordinate system, are introduced. While the world coordinate system is fixed, the source coordinate system rotates simultaneously with the data acquisition system. Since the definition of the source coordinate system changes with the projection sampling geometry, this coordinate system is introduced separately for each geometry. The underlying coordinate systems are used to give a mathematical description for the calculation of a line integral that is either called the Radon transform for the parallel-beam geometry or the divergent fan-beam transform for the fan-beam geometry. More detailed information about the data acquisition process may be found in several textbooks, e.g., in Herman [Herm 09], Kak and Slaney [Kak 01], Hsieh [Hsie 09], Buzug [Buzu 08], and Natterer [Natt 86].

3.1 Scanner Geometry

The description of scanner geometry involves several coordinate systems: a world coordinate system, a detector coordinate system, an image coordinate system, and a source coordinate system. The source coordinate system has a specific definition for each sampling geometry. Only the description of the source coordinate system in parallel-beam and in fan beam geometry for two and three dimensions is explained in Section 3.2 to Section 3.4 together with the calculation of the X-ray line integrals. By definition, all coordinate systems have the same origin.

World coordinate system. The world coordinate system in three dimensions is defined by the orthogonal unit vectors $e_x = [1, 0, 0]^T$, $e_y = [0, 1, 0]^T$, and $e_z = [0, 0, 1]^T$. In two dimensions, the orthogonal unit vectors are given by $e_x = [1, 0]^T$ and $e_y = [0, 1]^T$, respectively. This coordinate system is fixed during a scan.

Detector coordinate system. In modern CT scanners the detector consists of a 2D array of detector elements with N_u rows and N_v columns. Let Δu and Δv be the detector pixel width and height, respectively. Let the positive parameters u_{off} and v_{off} define the offset of the detector in detector element units. Then, for example a desired detector position $(u_i, v_j)^T$ is defined by

$$u_i = (-u_{\text{off}} + i)\,\Delta u \qquad \text{with} \qquad i = 0, 1, \ldots, (N_u - 1)\,, \qquad (3.1)$$
$$v_j = (-v_{\text{off}} + j)\,\Delta v \qquad \text{with} \qquad j = 0, 1, \ldots, (N_v - 1)\,. \qquad (3.2)$$

The vector from the center of the detector coordinate system to a specific detector pixel (u_i, v_j) is defined as $\boldsymbol{u}_{i,j}$.

Image coordinate system. Points within the FOV are identified by Cartesian coordinates, x, y and z. The final reconstruction is represented on a finite set of points called (x_k, y_l, z_m), where k, l and m are positive indices. These points are uniformly distributed in x, y and z and the number of samples in x is N_x, in y is N_y and that in z is N_z. The image pixel size is denoted as $\Delta x, \Delta y, \Delta z$. The positive parameters x_{off}, y_{off} and z_{off} define the offset of the image in image pixel units. Each image pixel can be associated with a point (x_k, y_l, z_m) with

$$x_k = (-x_{\text{off}} + k)\,\Delta x \qquad \text{with} \qquad k = 0, 1, \ldots, (N_x - 1)\,, \qquad (3.3)$$
$$y_l = (-y_{\text{off}} + l)\,\Delta y \qquad \text{with} \qquad l = 0, 1, \ldots, (N_y - 1)\,, \qquad (3.4)$$
$$z_m = (-z_{\text{off}} + m)\,\Delta z \qquad \text{with} \qquad m = 0, 1, \ldots, (N_z - 1)\,. \qquad (3.5)$$

This pixel is naturally centered on (x_k, y_l, z_m). The samples obtained by varying k while keeping l and m fixed are said to form a row of samples. Similarly, the samples obtained by varying l while keeping k and m fixed are said to form a column of samples. The vector from the center of the image coordinate system to a specific image pixel (x_k, y_l, z_m) is defined as $\boldsymbol{x}_{k,l,m}$.

When the sampling distance in x and y is the same, i.e., $\Delta x = \Delta y$, then this distance is denoted as Δ_{xy}. The same concept applies for three dimensions, where Δ_{xyz} refers to an equivalent sampling distance in all dimensions.

3.2   Two-Dimensional Radon Transform

The data acquisition system in parallel-beam geometry in two dimensions is defined by $\boldsymbol{\theta} = [\cos\theta, \sin\theta]^T$ and $\boldsymbol{\theta}^{\perp} = [-\sin\theta, \cos\theta]^T$, where θ is the projection angle. The

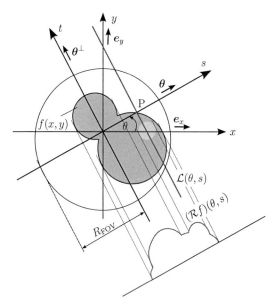

Figure 3.1: Definition of the fixed (x, y) coordinate system and of the rotating (s, t) coordinate system. The source coordinate system represents the data acquisition system and is defined by the orthogonal unit vectors $\boldsymbol{\theta}$ and $\boldsymbol{\theta}^{\perp}$. The thick gray line $\mathcal{L}(\theta, s)$ represents the integration path of the Radon transform, $(\mathcal{R}f)(\theta, s)$. All parallel-beam projections are obtained by shifting $\mathcal{L}$ along the s axis. R_{FOV} denotes the FOV radius.

projection angle is measured relative to the x-axis. The unit vectors $\boldsymbol{\theta}$ and $\boldsymbol{\theta}^{\perp}$ are perpendicular to each other and are associated with Cartesian coordinates, s and t, measured along $\boldsymbol{\theta}$ and $\boldsymbol{\theta}^{\perp}$, respectively.

Any line, $\mathcal{L}$, along an X-ray beam is defined by θ and s and is parallel to $\boldsymbol{\theta}^{\perp}$. Furthermore, any vector may be expressed in the (x, y) coordinate system by $x\boldsymbol{e}_x + y\boldsymbol{e}_y$ or in the (s, t) coordinate system using the linear combination $s\boldsymbol{\theta} + t\boldsymbol{\theta}^{\perp}$. The interrelations between the world and the source coordinate system are illustrated in Figure 3.1.

Let $f(\boldsymbol{x})$ be a function expressing the attenuation coefficients of an object in the FOV at $\boldsymbol{x} = x\boldsymbol{e}_x + y\boldsymbol{e}_y = s\boldsymbol{\theta} + t\boldsymbol{\theta}^{\perp}$. The function $f(\boldsymbol{x})$ is zero outside of the object. Then, the Radon transform, $\mathcal{R}$, of the function f is given by

$$g_p(\theta, s) = (\mathcal{R}f)(\theta, s) := \int_{-\infty}^{\infty} f(s\boldsymbol{\theta} + t\boldsymbol{\theta}^{\perp})\, \mathrm{d}t\,. \tag{3.6}$$

Hence, the result of $g_p(\theta, s)$ represents a measurement of a single line integral along $\mathcal{L}(\theta, s)$, where the line is parallel to $\boldsymbol{\theta}^{\perp}$ at a signed distance s, and where the data acquisition system is rotated by an angle θ. For a fixed angle θ, a parallel-beam

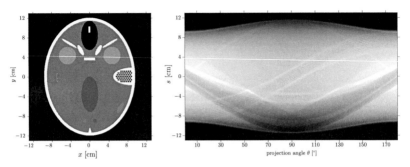

Figure 3.2: The FORBILD head phantom (display window: $[0, 100]$ HU) *(left)* with the corresponding Radon transform $(\mathcal{R}f)(\theta, s)$ *(right)*.

projection of f is created by accumulating the measurements while shifting $\mathcal{L}$ along the s axis.

Due to the limited FOV radius, R_{FOV}, the 2D Radon transform only needs to be calculated for $s \in [-R_{\mathrm{FOV}}, R_{\mathrm{FOV}}]$. Two important properties of the Radon transform, the parity $(\mathcal{R}f)(\theta, s) = (\mathcal{R}f)(\theta + \pi, -s)$ and the 2π periodicity $(\mathcal{R}f)(\theta, s) = (\mathcal{R}f)(\theta + 2\pi m, s)$ for any integer m allow to further reduce the interval for θ to $\theta \in [0, \pi)$. This means, the Radon transform $(\mathcal{R}f)(\theta, s)$ is completely defined for $\theta \times s \in [0, \pi) \times [-R_{\mathrm{FOV}}, R_{\mathrm{FOV}}]$. The Radon transform $(\mathcal{R}f)(\theta, s)$ is also called a sinogram of $f(\boldsymbol{x})$.

Figure 3.2 shows the FORBILD head phantom that was developed at the former IMP chair at the FAU[1] [Laur] as an example for a function $f(\boldsymbol{x})$ together with the corresponding sinogram [Yu 12].

3.3   Divergent Beam Transform in 2D

The source coordinate system for the two-dimensional fan-beam geometry is spanned by the orthogonal unit vectors $\boldsymbol{e}_u$ and $\boldsymbol{e}_w$, respectively. Let λ define the polar angle of the source position with $\lambda \in [0, 2\pi)$. Then, the orthogonal unit vectors are defined as follows:

$$\boldsymbol{e}_u(\lambda) = [-\sin\lambda, \cos\lambda]^T , \tag{3.7}$$

$$\boldsymbol{e}_w(\lambda) = [\ \cos\lambda, \sin\lambda]^T . \tag{3.8}$$

The source trajectory is a circle with radius R_F and is also known as vertex path. The location of the source on the trajectory is called the vertex point and is defined as:

$$\boldsymbol{a}(\lambda) = [R_F \cos\lambda, \ R_F \sin\lambda]^T . \tag{3.9}$$

In fan-beam geometry, the X-ray beam starts at the source position and diverges in the direction the detector. The directions of the lines, $\boldsymbol{\alpha}(\lambda, \gamma)$, are generally

---

[1]http://www.imp.uni-erlangen.de/phantoms

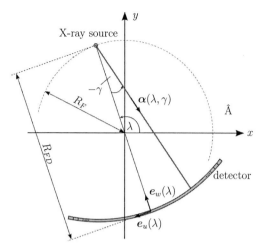

Figure 3.3: Source coordinate system for the two-dimensional fan-beam geometry with the associated parameters. R_F denotes the radius of the circular source trajectory and R_{FD} denotes the distance from the source to the detector. The thick gray line represents the integration path $\boldsymbol{a}(\lambda) + t\boldsymbol{\alpha}(\lambda,\gamma)$ with $t \in [0, R_{FD}]$.

parameterized using the the detector array that is in place. For a curved detector the detector elements are arranged on a circle segment of radius R_{FD} around the vertex point. The directions $\boldsymbol{\alpha}$ are best realized using equiangular rays defined by an angle γ, where γ is defined as the clockwise counted angle between the ray and the line from the source through the origin ($\boldsymbol{x} = (0,0)$) as it is shown in Figure 3.3. Then, the directions are given by

$$\boldsymbol{\alpha}(\lambda,\gamma) = -\cos\gamma\,\boldsymbol{e}_w + \sin\gamma\,\boldsymbol{e}_u\ . \qquad (3.10)$$

Let $f(\boldsymbol{x})$ be again a function expressing the attenuation coefficients of an object in the FOV. Note that any point along the X-ray beam is defined as $\boldsymbol{a}(\lambda) + t\,\boldsymbol{\alpha}(\lambda,\gamma)$ where t is a positive parameter. Then, the 2D divergent beam transform, $\mathcal{D}$, of the function f, also denoted as $g_f(\lambda,\gamma)$, is given by [Kak 01, Noo 04]

$$g_f(\lambda,\gamma) = (\mathcal{D}f)(\lambda,\gamma) := \int_0^\infty f(\boldsymbol{a}(\lambda) - t\cos\gamma\boldsymbol{e}_w + t\sin\gamma\boldsymbol{e}_u)\,\mathrm{d}t\ . \qquad (3.11)$$

Due to the limited FOV radius, the divergent beam transform only needs to be calculated within a certain maximal fan angle γ, i.e., for $\gamma \in [-\gamma_{max}, \gamma_{max}]$. The angle γ_{max} is related to both the FOV radius and the radius of the source and can be obtained by solving the following equation

$$R_{\mathrm{FOV}} = R_F \sin\gamma_{max}\ . \qquad (3.12)$$

Equation 3.11 can also be rewritten as a function of $(\mathcal{D}f)(\lambda, \boldsymbol{\alpha})$. Later in the text, both expressions $(\mathcal{D}f)(\lambda, \gamma)$ and $(\mathcal{D}f)(\lambda, \boldsymbol{\alpha})$ are used.

Even though current CT scanners collect data from a divergent beam, most reconstruction algorithms are based on parallel-beam data. The procedure of translating two-dimensional fan-beam data to parallel-beam data is referred to as parallel-beam rebinning. The link between (λ, γ) and (θ, s) is given by

$$\theta = \lambda + \frac{\pi}{2} - \gamma \ , \tag{3.13}$$

$$s = R_F \sin \gamma \ . \tag{3.14}$$

Note that the equations above will change when a possible FFS feature is used. Then, the expression of the circular scanning path (Eq. 3.9) becomes

$$\boldsymbol{a} = R_F \, \boldsymbol{e}_w(\lambda) + \delta(\lambda) \, \boldsymbol{e}_u(\lambda) \ , \tag{3.15}$$

with $\delta(\lambda) \neq 0$.

3.4   Divergent Beam Transform in 3D

The source coordinate system in three dimensional cone-beam geometry is spanned by the orthogonal unit vectors $\boldsymbol{e}_u$, $\boldsymbol{e}_v$, and $\boldsymbol{e}_w$:

$$\boldsymbol{e}_u = [- \sin \lambda, \cos \lambda, 0]^T \ , \tag{3.16}$$

$$\boldsymbol{e}_v = [\quad 0, \quad 0, 1]^T \ , \tag{3.17}$$

$$\boldsymbol{e}_w = [\quad \cos \lambda, \sin \lambda, 0]^T \ . \tag{3.18}$$

By definition, the vectors $\boldsymbol{e}_u$ and $\boldsymbol{e}_v$, respectively, are parallel to the direction of the detector rows and columns. The unit vector $\boldsymbol{e}_w$ points from the detector towards the origin to the X-ray source.

In three dimensions, the source trajectory is either circular or helical. For both vertex path descriptions, additional parameters are needed. Let λ_0 be the start angle, let z_0 be the position $z = z_0$ of the source in the plane at polar angle λ_0 and let p be the pitch between two points on the helix trajectory after a rotation of 2π. Then, the generalized source trajectory, $\boldsymbol{a}(\lambda)$, is defined as

$$\boldsymbol{a}(\lambda) = [R_F \cos(\lambda + \lambda_0), \ R_F \sin(\lambda + \lambda_0), \ z_0 + p \, \lambda/(2\pi)]^T \ . \tag{3.19}$$

For a circular CT scan, the pitch p in Equation 3.19 is equal to zero whereas for a spiral scan $p \neq 0$. Any line that connects the vertex point $\boldsymbol{a}(\lambda)$ with a given location on the detector (u, v) is of direction

$$\boldsymbol{\alpha}(\lambda(u, v), \gamma(u, v)) = \frac{-R_{FD} \cos \gamma \, \boldsymbol{e}_u + R_{FD} \sin \gamma \, \boldsymbol{e}_u + v \boldsymbol{e}_w}{\sqrt{R_{FD}^2 + v^2}} \ . \tag{3.20}$$

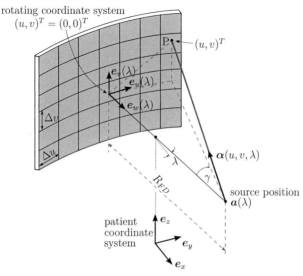

Figure 3.4: Source coordinate system for the three-dimensional cone-beam geometry with the associated parameters. The thick gray line represents the integration path $\boldsymbol{a}(\lambda) + t\boldsymbol{\alpha}(\lambda, \gamma, v)$ with $t \in [0, \sqrt{R_{FD}^2 + v^2}]$.

As in the two-dimensional case, any point along the X-ray beam is defined as $\boldsymbol{a}(\lambda) + t\,\boldsymbol{\alpha}(\lambda, \gamma, v)$, where t is a positive parameter. Then, the 3D divergent cone-beam transform, $g_c(\lambda, \gamma, v)$, is given by

$$g_c(\lambda, \gamma, v) = (\mathcal{D}f)(\lambda, \gamma, v) := \int_0^\infty f\left(\boldsymbol{a}(\lambda) + t\,\boldsymbol{\alpha}(\lambda, \gamma, v)\right)\,\mathrm{d}t. \qquad (3.21)$$

The geometry of the 3D divergent fan-beam transform is shown in Figure 3.4. Any point P on the detector can be specified using the angle γ and the coordinate v. Throughout this dissertation, the discussions are restricted to a data acquisition system with a vertex path following a circular CT scan.

Iterative Reconstruction Techniques

The reconstruction algorithms used over the last several decades were primarily analytical in nature. The most popular analytical reconstruction algorithm is the filtered back projection (FBP) algorithm, which follows the routine of weighting, filtering, and backprojecting each projection finally to receive a fully reconstructed image. Analytical reconstruction methods are generally very fast, especially on dedicated hardware. However, the question persists whether the reconstructed image quality (IQ) can be further improved using other classes of reconstruction algorithms such as iterative reconstruction (IR). Whereas analytical reconstruction algorithms are generally based on simplified assumptions, such as pencil beams representing line integrals, IR methods allow the modeling of the scanner's optics and physics. Furthermore, IR algorithms are said to enable CT exams with lower dose due to their general (statistical) approach and their data modeling options.

This chapter provides the basics of IR techniques. The introduction of IR algorithms is confined to the relevant parts needed throughout the dissertation. A description of analytical reconstruction algorithms is not given. An extensive overview of the theory of reconstruction algorithms and their details can be found in numerous articles [Feld 84, Kats 03, Floh 03, Stie 04, Noo 07] and in several textbooks [Kale 06, Buzu 08, Hsie 09, Herm 09, Geng 10, Doss 13]. Over the last years, significant effort was spent on the development of statistical iterative reconstruction algorithms for CT imaging. A small selection of two well-known iterative reconstruction algorithms is given in Section 4.3.

4.1   General Approach to Iterative Reconstruction

In contrast to the analytical calculation of a line integral $\mathcal{L}$ as described in Chapter 3, in which the object is sampled with a infinitely small line, IR methods allow a user specific modeling of the properties of that line integral. As an example, a line integral that is passing through tissue may consider the contribution of each pixel that is to be reconstructed is passed through by the line $\mathcal{L}$ differently. Thus, weights are introduced to reflect the contribution of the pixels to the X-ray beam. Taking all pixels along the line into account gives a full description of the calculation of the projection data.

Let $\boldsymbol{g} \in \mathbb{R}^M$ represent the projection data (sinogram) with $M = N_u \cdot N_v \cdot N_{\text{proj}}$, where N_{proj} is the total number of projections. Each component g_r in $\boldsymbol{g} = (g_1, \ldots, g_M)^T$ represents the projection data of the r-th line integral. Let the estimate of linear X-ray attenuation coefficients be represented by vectors $\boldsymbol{f} \in \mathbb{R}^{N_p}$ with $N_p = N_x \cdot N_y \cdot N_z$, where each component f_s in $\boldsymbol{f} = (f_1, \ldots, f_{N_p})^T$ represents a certain voxel of the reconstructed volume. Further, let $\mathbf{A} \in \mathbb{R}^{M \times N_p}$ be the system (projection) matrix [Toft 96] including the weights a_{rs} that reflect the contribution of the s-th voxel to the r-th line integral, which is represented by

$$g_r = \sum_{s=1}^{N_p} a_{rs} f_s \ . \tag{4.1}$$

Then, the complete set of projection data is obtained by the system of linear equations

$$\boldsymbol{g} = \mathbf{A}\boldsymbol{f} \ . \tag{4.2}$$

The matrix entries a_{ij} in the system matrix $\mathbf{A}$ strongly depend on the image representation technique. Different types of modeling techniques exist. Two commonly used techniques to represent the image with a finite number of unknowns are the sampling approach and the basis function approach. In the sampling approach, the image is represented by its values at a fixed number of locations that are typically equidistantly distributed in the direction of Cartesian coordinates [Buzu 08]. A popular sampling approach is the method of Joseph [Jose 82].

In the basis function approach, the image is represented by a finite linear combination of specific functions that are often selected as scaled and translated versions of a single function, called the mother function. Popular mother functions include the blobs [Lewi 90, Mate 96] and the B-splines [Horb 02]. Siddon's well-known method [Sidd 85] is a basis function approach using the square pixel as mother function, i.e., the B-splines of order zero. Another very popular method, the distance-driven method [De M 02], is obtained by calculating a strip integral instead of a line integral. More basis function approaches can be found in Peters [Pete 81], Noo et al. [Noo 12], and Schmitt et al. [Schm 12a]. These old and new approaches may be fast and straightforward; however, some of them struggle with reconstruction artifacts. Such artifacts may be due to aliasing of neighboring DC-components (DC-aliasing) [Dani 03] or to the fact that the area under the basis functions do not sum to one. A detailed description of some representation techniques used throughout this dissertation is given in Section 6.2.

As most of the pixels in the volume do not contribute to the projection data, many weights a_{rs} are equal to zero which means that the system matrix $\mathbf{A}$ is sparse. This is schematically shown in Figure 6.1 (p. 55) for Joseph's method as described in Section 6.2.1. Both the sparsity of $\mathbf{A}$ and the value of the matrix entries a_{rs} highly depend on the individual choice of the image representation technique and the chosen model for the CT measurement: line or strip integral.

In two and three dimensions, a duality exists between the analytical expression and the previously described approach of calculating the line integrals. For instance, in 2D the duality is identified by

$$
\begin{array}{ccc}
g_p = & (\mathcal{R}f) \\
\updownarrow & \updownarrow\updownarrow & \\
g = & \mathbf{A}\,\boldsymbol{f}\,.
\end{array}
\tag{4.3}
$$

Both equations give a description of the data acquisition process. The first equation can be solved analytically. For example, a solution of the analytical 2D problem (Eq. 3.6) may be obtained using FBP [Kak 01, Buzu 08] or in 3D (Eq. 3.21) using the algorithm proposed by Feldkamp, David and Kress (FDK algorithm) [Feld 84]. While analytical equations are sometimes difficult to solve, the solution of Equation 4.2 is obtained using iterative reconstruction methods since directly solving Equation 4.2 turns out to be mathematically difficult for several reasons. First, exactly solving the system of linear equations requires data that is taken under idealized physical conditions, but real measurements are afflicted with noise. Second, in a CT scanner the number of projections is usually higher than the number of pixels that are to be reconstructed ($M > N_P$), leading to an over-determined system of equations. Furthermore, $\mathbf{A}$ is usually very large and sparse, contains very small singular values which affect stability of its inversion, and does not have a simple structure [Buzu 08].

4.2   Framework of Iterative Reconstruction Methods

Most iterative reconstruction methods are based on minimizing or maximizing an objective function, $\Phi(\boldsymbol{f},\boldsymbol{g})$, also called cost function. Generally, the objective function is given by a sum of two terms. The first term, $\Phi_M(\boldsymbol{f},\boldsymbol{g})$, includes the measurement statistics and the geometry of the data acquisition process. The second term, $\Phi_R(\boldsymbol{f})$, is a non-linear regularization term that mitigates noise and artifacts. Its definition is completely independent of the measurement process. The general expression of the objective function is defined as

$$
\Phi(\boldsymbol{f},\boldsymbol{g}) = \Phi_M(\boldsymbol{f},\boldsymbol{g}) + \Phi_R(\boldsymbol{f})\,.
\tag{4.4}
$$

Depending on the optimization strategy of the objective function, iterative reconstruction algorithms are subsequently divided into two categories: i) non-constrained iterative image reconstruction; and, ii) constrained iterative image reconstruction.

4.2.1   Non-Constrained Iterative Image Reconstruction

A non-constrained iterative image reconstruction seeks the solution, $\boldsymbol{f}^*$, of Equation 4.2 without any additional constraints on the image, i.e., $\Phi_R = 0$. A generalized inverse[1] of Equation 4.2 is obtained by solving a least squares problem that minimizes the Euclidean distance between the forward projected estimate, $\mathbf{A}\boldsymbol{f}$, and the

---

[1]For a typical imaging problem, the matrix $\mathbf{A}$ is not square.

measured projection data, g. The Euclidean distance is expressed in the data fidelity term of the cost function, Φ_M. Formally, the unconstrained iterative reconstruction problem can be written as

$$f^* = \arg\min_f \Phi_M(f, g) \ , \qquad \text{where} \qquad (4.5)$$

$$\Phi_M(f, g) = \|Af - g\|_2^2 \ . \qquad (4.6)$$

The expression in Equation 4.6 is also referred to as a non-weighted least squares difference. As Φ_M is a convex function there exists always a solution for this optimization problem. The solution is called the least squared minimum norm solution [Buzu 08].

In some situations, a constant diagonal ray weighting matrix[2], W, that includes the measurement statistics, for instance, the statistical noise model, is desired. Then, Φ_M is expressed as a weighted least squares difference, given as

$$\Phi_M = \|W(Af - g)\|_2^2 \ . \qquad (4.7)$$

Let $\tilde{A} = WA$ and $\tilde{g} = Wg$, then Equation 4.7 may also be expressed as

$$\Phi_M = \left\|\tilde{A}f - \tilde{g}\right\|_2^2 \ , \qquad (4.8)$$

and therefore has conceptually the same form as Equation 4.6. Below, to be as general as possible, each reconstruction algorithm is described using the expression of Φ_M as it is given in Equation 4.8. Note that, if no statistical weighting matrix is used, W can be interpreted as the identity matrix, I.

4.2.2   Constrained Iterative Image Reconstruction

Often, iterative image reconstruction methods tend to be unstable. This is reflected, e.g., in the fact that oscillations in the search for optimal parameters f^* may occur or that the iterates get more and more distorted by high frequency artifacts and noise as the number of iterations increase. Both effects are caused by the ill-posedness of the tomographic problem[3] and insufficient restrictions on the space of possible solutions. In CT, the measurement process suppresses high frequencies fairly sufficiently. However, the few high frequency components will quickly become dominant when noise is added. Thus, strong amplification of these noise components may appear, which leads to unusual reconstruction results. More details about ill-posed problems can be found in Hansen [Hans 98] and Jiang and Wang [Jian 03].

Therefore, a regularization term is introduced to eliminate such undesired effects and to restrict the solution space to certain well-behaved functions. Due to better conditioning of the inverse problem, this may also lead to a faster convergence of the

---

[2]This assumption is correct if the measurements are assumed to be statistically independent, which is usually a good approximation for CT imaging.

[3]A problem is said to be ill-posed if its solution is not unique, or if the solution does not continuously depend on the input data, i.e., if a small perturbation of input data may lead to a large perturbation of the solution [Hans 98, Jian 03, Sunn 09].

algorithm. In the context of iterative reconstruction, several regularization techniques exist in the literature, two of which are described below.

A very popular regularization method is based on minimizing the cost function that includes a regularization term which is non-zero, i.e., $\Phi_R \neq 0$. Then, the basic formulation of a constrained iterative image reconstruction problem is defined as

$$\boldsymbol{f}^* = \arg\min_{\boldsymbol{f}} \Phi(\boldsymbol{f}, \boldsymbol{g}) , \qquad \text{where} \qquad (4.9)$$

$$\Phi(\boldsymbol{f}, \boldsymbol{g}) = \Phi_M(\boldsymbol{f}, \boldsymbol{g}) + \Phi_R(\boldsymbol{f}) . \qquad (4.10)$$

The design of the regularization term Φ_R is usually chosen to penalize certain structures and properties that are less likely to appear in the image than others. Many different non-convex and convex potential functions have been suggested in the literature. In CT, convex potential functions are usually preferred due to the convexity of the measurement data term Φ_M. The sum of two convex functions is a convex function whereby always one minima consists. Thus, a convex potential function has the advantage that the objective function becomes also convex.

The strength of Φ_R is controlled by the positive regularization parameter, β_R, while the design of the potential function, $\phi(f)$, affects the modality as the reconstructions are penalized. Let ω_{kl} be the distance between voxel k and l within some small neighborhood, then the expression of the regularization term in its generally form is

$$\Phi_R(\boldsymbol{f}) = \beta_R \sum_k \sum_l \omega_{kl}^{-1} \, \phi(f_l - f_k) . \qquad (4.11)$$

Examples of convex potential functions are the Huber penalty [Hube 11], the Gibbs prior [Boum 93], or the generalized Gaussian Markov random field [Thib 07]. Throughout this dissertation, the focus was on the convex Fair potential function [Fair 74, Rama 12],

$$\phi^{\mathrm{FP}}(x) = \epsilon \cdot \left[\left| \frac{x}{\epsilon} \right| - \log\left(1 + \left| \frac{x}{\epsilon} \right| \right) \right] , \qquad (4.12)$$

which is a smooth edge-preserving regularization, and where ϵ is a positive parameter and on the quadratic potential function

$$\phi^{\mathrm{QP}}(x) = \left(\frac{x^2}{2\,\epsilon} \right) . \qquad (4.13)$$

Both regularization functions are visualized in Figure 4.1.

Another popular method of regularization is to stop the reconstruction algorithm after a finite number of iteration steps [Vekl 87, Hebe 88]. This regularization method is theoretically unsatisfactory since the reconstruction algorithm is stopped before the final convergence point has been reached. This means, the original mathematical problem formulation and the result do not match. Finding a satisfactory stopping criterion is not obvious; it needs to be defined with respect to the particular optimization method. The lack of some feasibility parameters will be further discussed in Chapter 7.

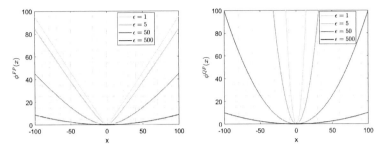

Figure 4.1: Illustration of the Fair potential *(left)* and the quadratic potential function *(right)*.

4.3 Iterative Reconstruction Algorithms

Iterative reconstruction algorithms are an important research topic in CT imaging. Thus, many different iterative reconstruction algorithms exist in the literature. Each specific iterative reconstruction technique is based on an error matrix that is used to determine the update from one iteration to the next. The goal is to reduce the error in subsequent iterations. Below, the iterative schema of two different algorithms is described. While the Landweber algorithm provides a solution for a non-constrained image reconstruction problem, the iterative coordinate descent method may also be used to solve a constrained image reconstruction problem.

4.3.1 Landweber Method

Throughout this dissertation, the Landweber algorithm [Land 51] was used to estimate the minimum-norm minimizer of a non-constrained image reconstruction problem as defined in Equation 4.5. Conceptually, the updating formula in the Landweber algorithm of the (n)-th to the $(n+1)$-th iteration can be interpreted as a steepest gradient descent method. For a weighted least squares difference, the Landweber iteration schema is given by

$$\boldsymbol{f}^{(n+1)} = \boldsymbol{f}^{(n)} + \eta \, \tilde{\mathbf{A}}^T \left(\tilde{\boldsymbol{g}} - \tilde{\mathbf{A}} \boldsymbol{f}^{(n)} \right) \, , \tag{4.14}$$

where the parameter η controls both the convergence and the step size of the Landweber update in each iteration. A proof of that and of the limits of η is given in the appendix A.1.

The procedure of each Landweber iteration can be illustrated as follows: The projection matrix $\tilde{\mathbf{A}}$ maps the estimate $\boldsymbol{f}^{(n)}$ onto $\tilde{\mathbf{A}}\boldsymbol{f}^{(n)}$, the forward projection of the estimate. The error matrix, $\tilde{\boldsymbol{g}} - \tilde{\mathbf{A}}\boldsymbol{f}^{(n)}$, is used to determine a new estimate by first backprojection the error matrix with the adjoint operator $\tilde{\mathbf{A}}^T$. The result is further multiplied with η, and finally added to the old estimate $\boldsymbol{f}^{(n)}$ to obtain the new estimate $\boldsymbol{f}^{(n+1)}$. Note that in the Landweber algorithm each component of $\boldsymbol{f}$ is simultaneously updated in every Landweber iteration step.

Algorithm 4.1: Landweber algorithm

Input: W, g, $f^{(0)}$, choice of the forward projection model[a]
Output: reconstructed image volume f

// pre-computations before starting with the first iteration
1   compute projector elements a_{rs}
2   apply the statistical weights, i.e., compute: $\tilde{\mathbf{A}} = \mathbf{W}\mathbf{A}$ and $\tilde{g} = \mathbf{W}g$

// Landweber iterations
3   **for** $n = 1 \dots N_{it}$ **do**   (iteration number)
4    $\quad$ compute forward projection estimate: $\tilde{\mathbf{A}}f^{(n)}$
5    $\quad$ compute difference: $\Delta g := (\tilde{g} - \tilde{\mathbf{A}}f^{(n)})$
6    $\quad$ compute new estimate by backprojecting the difference: $\tilde{\mathbf{A}}^T \Delta g$
7    $\quad$ multiplying the previous result with the step length η
8    $\quad$ update image estimate: f^{n+1}
9    $\quad$ *repeat the previous steps until convergence is achieved*[b]
10   **end**

---

[a] The following forward projection models have been implemented: Joseph's method, distance-driven method, and B-splines (see Sec. 6.2);
[b] The exact stopping criterion is defined by the user.

Ideally, that procedure is repeated over and over until the assumed and measured values are the same. However, in most cases the fixed point of the Landweber algorithm is not desirable, since resolution and noise in the reconstructions are progressively traded for one another as the number of iterations increase. Thus, the Landweber algorithm is often stopped after an acceptable limit of reconstruction is received. Therefore, throughout the dissertation the Landweber algorithm was always stopped after a fixed number of iterations. Unless otherwise stated, 1000 iterations were performed. A detailed overview of the single steps of the Landweber algorithm is shown in Algorithm 4.1.

The convergence of the Landweber algorithm is primarily determined by the parameter η. Convergence is only guaranteed if η is between $0 < \eta < 2/\sigma_{\max}^2$, where $\sigma_{\max}$ is the maximum singular value of the projection matrix $\tilde{\mathbf{A}}$. A stable approach of estimating $\sigma_{\max}$ is the Power method [Golu96]. The Power method iteratively determines the maximum singular value. That approach is stable and comparatively fast. Unless otherwise stated, the estimate of $\sigma_{\max}$ after 5 iterations was used for further calculations. Additionally, to avoid high variations from one iteration to the next, the step size was always chosen to be $\eta = 0.90 \cdot 2/\sigma_{\max}^2$. Large variations may occur when η is set too close to its upper limit of $2/\sigma_{\max}^2$.

If the Landweber algorithm is not stopped and if η was chosen correctly, the Landweber algorithm converges theoretically to the fixed point $f^* = (\tilde{\mathbf{A}}^T \tilde{\mathbf{A}})^{-1} \tilde{\mathbf{A}}^T \tilde{g}$ (see also appendix A.1).

4.3.2   Iterative Coordinate Descent Method

The iterative coordinate descent (ICD) method was first applied by Sauer and Bouman to least squares Bayesian tomographic reconstruction in 1993 [Saue 93]. The method has become a well-known approach to calculate the minimizer of a strictly convex constrained image reconstruction problem [Boum 96, Thib 00, Zhen 00, Thib 07, Bens 10]. The ICD method benefits from a fast convergence of high frequencies and from the relatively low computational requirements. More information about the convergence behavior can be found in Abatzoglou and O'Donnell [Abat 82] and Luo and Tseng [Luo 92].

The expression of Φ_M (Eq. 4.8) in conjunction with the general expression for the regularization function Φ_R (Eq. 4.11) may be used to formulate an approximate image estimate $\boldsymbol{f}^*$:

$$\boldsymbol{f}^* = \arg\min_{\boldsymbol{f}} \left\{ (\tilde{\mathbf{A}}\boldsymbol{f} - \tilde{\boldsymbol{g}})^T (\tilde{\mathbf{A}}\boldsymbol{f} - \tilde{\boldsymbol{g}}) + \beta_R \sum_k \sum_l \omega_{kl}^{-1} \, \phi(f_l - f_k) \right\} . \qquad (4.15)$$

The ICD approach to the solution is based on a voxel-by-voxel updating technique. This means, the components of $\boldsymbol{f}$ are updated one after the other, which corresponds to a local minimization of the objective function. During the optimization of the component f_s of $\boldsymbol{f}$ all other voxel values are kept unchanged. Under the assumption that the components of $\boldsymbol{f}$ are updated one after another, the optimization problem of the component f_s is

$$f_s^{(n+1)} = \arg\min_{f_s} \left\{ \Phi_s \left(f_1^{(n+1)}, \ldots, f_{s-1}^{(n+1)}, f_s, f_{s+1}^{(n)}, \ldots, f_{N_p}^{(n)} \right) \right\} . \qquad (4.16)$$

Note that the components $f_1, \ldots, f_{s-1}$ have already been updated; then, f_s is updated next, followed by the components $f_{s+1}, \ldots, f_{N_p}$. After a component has been updated, the iteration counter is incremented by one. A *full* iteration is achieved when all components have been updated exactly once. Hence, a full iteration contains N_p one-dimensional optimization steps that minimize Φ_s successively. Given that $\tilde{\boldsymbol{a}}_{r*}$ is the r-th row of $\tilde{\mathbf{A}}$ the one-dimensional cost function, Φ_s, in Equation 4.16 is given by

$$\Phi_s = \sum_{r=1}^M \left(\tilde{a}_{rs} \left(f_s - f_s^{(n)} \right) + \tilde{\boldsymbol{a}}_{r*} \boldsymbol{f}^{(n)} - \tilde{g}_r \right)^2 + 2\beta_R \sum_k \omega_{ks}^{-1} \, \phi \left(f_s - f_k^{(n)} \right) . \qquad (4.17)$$

The minimum of Equation 4.17 is found by computing the root of the first derivative of Φ_s. While the first derivative is computed analytically, the derivative's root is found numerically. Let ϕ' be the first derivative of the prior potential and let

$$\Lambda_1 = \sum_{r=1}^M \tilde{a}_{rs} \left(\tilde{\boldsymbol{a}}_{r*} \boldsymbol{f}^{(n)} - \tilde{g}_r \right) , \qquad (4.18)$$

$$\Lambda_2 = \sum_{r=1}^M \tilde{a}_{rs}^2 , \qquad (4.19)$$

Algorithm 4.2: ICD algorithm

Input: $\mathbf{W}$, $\mathbf{g}$, $\boldsymbol{f}^{(0)} \neq 0$, choice of the forward projection model [a] and the regularization function [b]

Output: reconstructed image volume $\boldsymbol{f}$

// pre-computations before starting with the first iteration

1 compute projector elements a_{rs}

2 apply the statistical weights, i.e., compute: $\tilde{\mathbf{A}} = \mathbf{WA}$ and $\tilde{\boldsymbol{g}} = \mathbf{W}\boldsymbol{g}$

3 compute: $\tilde{\mathbf{A}}\boldsymbol{f}^{(0)}$

// ICD iterations

4 **for** $n = 1 \ldots N_{it}$ **do**   (iteration number - full ICD iterations)

5   **for** $s = 1 \ldots N_p$ **do**   (voxel scan pattern - single image space iterations)

6   compute Λ_1, Λ_2

7   perform half-line search to find the root of Equation 4.20

8   update selected voxel: $f_s^{(n+1)}$

9   update forward projection estimate:

        $\tilde{\mathbf{A}}\boldsymbol{f}^{(n+1)} = \tilde{\mathbf{A}}\boldsymbol{f}^{(n)} + \tilde{a}_{r*}(f_s^{(n+1)} - f_s^{(n)})$

10   **end**

11   *repeat the previous steps until convergence is achieved* [c]

12 **end**

---

[a] The following forward projection models have been implemented: Joseph's method, distance-driven method, and B-splines (see Sec. 6.2).

[b] The following potential functions and their corresponding derivations have been implemented: Fair potential and a quadratic potential.

[c] The exact stopping criterion is defined by the user.

---

then the new voxel value is found by solving

$$\Lambda_1 - \Lambda_2(f_s^{(n)} - f_s) + \frac{\beta_R}{2} \sum_k \omega_{ks}^{-1}\, \phi'(f_s - f_k^{(n)}) \bigg|_{f_s = f_s^{(n+1)}} = 0 \ . \tag{4.20}$$

A half-interval search to some tolerance around the root was performed to solve Equation 4.20, which showed a stable convergence.[4] The computational effort was relatively low. This is because the number of adjacent pixels is generally low, and that Λ_1 and Λ_2 were precomputed before starting the half-interval search.

As mentioned before, all voxel values were updated once the iteration counter is incremented by one. Usually, this process is repeated until convergence is achieved. A detailed overview of the single steps of the ICD method is shown in Algorithm 4.2.

The definition of convergence differs widely in the literature. To give just two out of many examples: i) stop iterating when the change for each voxels' value from one to the next iteration is smaller than 1 HU [Thib 07]; or, ii) stop iterating when the decrease of the objective function value is less than 0.001 [Fess 97].

---

[4] A first approach using the Newton-Raphson method to minimize Equation 4.20 showed some instabilities. Hence, the half-line search was preferred.

Note that only for reasons of clarity has the scan pattern been assumed to be one after the other in Equation 4.16. Generally, the scan pattern can be selected to be either one by one or randomly. The choice may influence the convergence speed, i.e., some publications showed that a random selection minimizes the correlation between adjacent updates and maximizes the convergence speed [Saue 93, Bows 98, Thib 00].

4.4   Strengths and Weaknesses of Iterative Reconstruction Methods

The most valuable advantage of IR techniques is that realistic models of the imaging system may be taken into account. Such models may include a detailed description of the measurement process, i.e., a model that views a detector pixel as an area instead of a single point, or a model that includes an extensive geometric ray profile description [Hofm 14]. Moreover, IR techniques have been found worthwhile in reducing both metal artifacts [De M 00] and beam hardening artifacts significantly, for example, by applying an appropriate multi-energetic X-ray model [De M 01, Elba 02]. Generally, IR algorithms suppress artifacts due to missing or incomplete data much better than analytical methods, which might be because IR methods make use of all available data. The mathematical approach of minimizing an objective function can effectively incorporate a statistical noise model as well as a regularization term that mitigates noise in the reconstructed images. In this way, IR techniques usually provide a more accurate reconstruction. This accuracy also could occur because solving a system of linear equations is usually easier than solving an integral equation. This in turn allows a reduction of the patient radiation dose while maintaining or improving IQ at the same time. With IR techniques, i.e., the average effective dose was 8.9 ± 7.1mSv and before IR it was 10.1 ± 7.8mSv [Noel 13].

However, IR techniques also have drawbacks. The main disadvantage is that IR algorithms are computationally expensive. This is also reflected in the high number of publications proposing new IR algorithms with the aim of optimizing the reconstruction time, and the number of iterations to obtain an IR technique that can be used in daily clinic routine. In order to speed up the reconstruction process, algorithms based on subsets of the projection dataset were proposed, called the ordered subset (OS) approach [Kamp 98, Erdo 99]. However, this approach is known to be unstable and non-convergent when the reconstruction parameters are not carefully selected. Yet, selecting or determine IR reconstruction parameters can be quite difficult. When defining an IR method, many decisions have to be made, e.g., the selection of the IR algorithm, the definition of the projection matrix, the definition of additional geometrical models, the selection of a possible regularization, etc. Most of the parameters directly influence the image reconstruction result. In addition to that, the reconstruction result might also depend on the initialization of the IR algorithm, particularly if the reconstruction is stopped after a given number of iterations. At the time of writing this dissertation and to the best knowledge of the author, a common accepted stopping criteria with respect to image quality does not exist. The last disadvantage that will be mentioned here is the lack of ability to reconstruct only a small local region of interest (LROI) [Zieg 08, Sunn 09].

Image Quality Assessment

Image quality (IQ) assessment is a central concept for evaluating the key performance parameters of a CT scanner, such as resolution or noise. A simple way to improve the image quality on a CT scanner is to increase the patient's radiation dose, which would additionally present important health concerns. Currently, the question of how the patient's radiation dose could be further minimized is of high interest. However, it is very important that the final reconstructed image be assessed with respect to how well it conveys all anatomical or functional information. The signs of a disease or injury need to be clearly visible such that the interpreting radiologist can make an accurate diagnosis. Different quantitative metrics exist to evaluate certain IQ properties, such as basic metrics and task-based metrics.

This chapter starts with a description of the mathematical test object, called phantom, that was used to evaluate image quality followed by a section that describes the quantitative evaluation using basic metrics. In this section, the definition of the physical parameters that are used to measure quantitative IQ parameters is given. Section 5.3 describes the quantitative evaluation using task-based metrics. A task-based image quality assessment may be based on mathematical model observers and on human observers, respectively. The chapter ends with a discussion that explains the weaknesses and strengths for the basic and task-based IQ assessment.

5.1 Phantom

Mathematical test objects have been proven to be very helpful for the development and evaluation of new reconstruction algorithm or for the assessment of image quality. Such phantoms usually include sophisticated details to challenge the reconstruction algorithm in example resolution patterns. The mathematical phantom that was used for the studies presented in this dissertation was the FORBILD head phantom [Laur], which is shown in Figure 3.2 together with the corresponding sinogram.

The phantom was proposed by a group of CT researchers at the former Institute of Medical Physics in Erlangen at FAU, Germany, together with scientists from Siemens

Healthcare GmbH. The main goal of this phantom is on the one hand to model different parts of the anatomy and on the other hand to be very challenging in order to improve the robustness and the reliability of evaluations. By design, the inner ear region consists of high-contrast objects that generate artifacts. However, the phantom still allows the evaluation of low-contrast object detectability. While, in example, the Shepp-Logan phantom [Shep 74] only consists of ellipsoids, the FORBILD head phantom is built using ellipsoids, cylinders, cones, or a portion thereof. In 2012, a full description of the 2D FORBILD head phantom together with associated Matlab simulation tools was published in Yu et al. [Yu 12]. That description was used for all 2D simulations throughout the dissertation.

5.2   Quantitative Evaluation: Basic Metrics

A quantitative evaluation using basic metrics includes conventional physical parameters such as resolution, bias, and noise properties. The use of basic metrics allows to assess IQ parameters very fast and objective. The main advantage of basic metrics is the ease of use and transferability to any reconstructed image.

5.2.1   Spatial Resolution

Spatial resolution is an important feature to evaluate the image quality, especially if various iterative reconstruction algorithms are compared against each other. A distinction is made according to whether the spatial resolution is measured in-plane (xy-plane) or orthogonal to that, cross-plane (z-direction) [Hsie 09]. Both measurements depend on several factors, such as, the focal spot size, the detector pixel size, the scanner geometry, the reconstruction algorithm, inter alia. These factors limit the maximally attainable spatial resolution of the system.

The in-plane spatial resolution is typically represented by modulation transfer functions (MTFs). In practice, a thin wire test object that is placed orthogonal to the xy-plane is first scanned and then reconstructed to a very fine grid. The thin wire approximately represents a Dirac line which means that the reconstructed axial image provides the axial point spread functions (PSFs). The MTF is then obtained by taking the Fourier transform of the PSF [Ross 69, Bush 11]. The MTF is measured in line pairs per centimeter (lp/cm) or in line pairs per millimeter (lp/mm).

When using iterative reconstruction algorithms, the spatial resolution normally varies from one image representation to the other and highly depends on the choices made with regard to the image representation. For that reason, the afore mentioned method is not appropriate. Instead, the method proposed by Li et al. [Li 07] was applied. This method is an extension of the edge-based MTF method published by Judy [Judy 76] and can be summarized in four steps: i) collecting contributing samples; ii) binning; iii) differentiation; and, iv) Fourier transform. By applying steps i to iv it is possible to measure resolution in any location, along any direction, for any shaped object, and without changing the sampling density. Each single step works as follows:

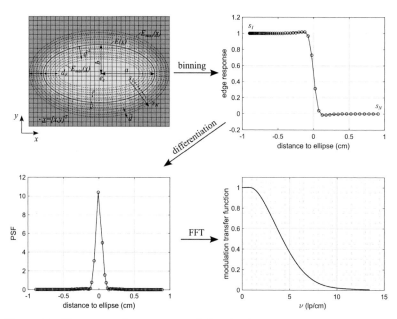

Figure 5.1: Illustration of steps i to iv of the edge-based MTF method. First, the contributing samples are collected *(top left)*. A binning step is performed to obtain the edge response function *(top right)*. The differentiation of the edge response returns the PSF *(bottom left)*. Finally, the MTF is obtained by Fourier transforming the PSF *(bottom right)*.

i) Collecting contributing samples. Let $f(\boldsymbol{x})$ be the reconstructed image of the test object, which is a solid ellipse with center $\boldsymbol{e}_c = [x_c, y_c]^T$, with a major axis a, with a minor axis b, and with ϕ_e describing the angle between the x-axis and the major axis of the ellipse. Let $E(\boldsymbol{x})$ be a function that describes the border of that ellipse, let $\boldsymbol{d}^{\perp}$ be the orthogonal unit vector to $E(\boldsymbol{x})$ defining the direction along the resolution is measured, and, let d_E be a predefined distance that is by definition half of the smallest value of a or b: $d_E = 0.5 \min(a, b)$. Furthermore, let $E_{\min}(\boldsymbol{x})$ and $E_{\max}(\boldsymbol{x})$ be scaled versions of the ellipse $E(\boldsymbol{x})$ such that those ellipses are located at a distance of $\pm d_E$ from $E(\boldsymbol{x})$. A graphic representation of the interrelations is shown in Figure 5.1 (top left).

For the next steps only the pixels between the curves $E_{\min}(\boldsymbol{x})$ and $E_{\max}(\boldsymbol{x})$ are considered since they already contain all information needed to calculate the MTF. For each point P between those two ellipses the distance between P and the ellipse $E(\boldsymbol{x})$ is calculated. That distance is achieved with the point Q on the ellipse $E(\boldsymbol{x})$ that is closest to P. If P is within the ellipse, the distance is given a negative sign, otherwise it is given a positive sign. The set of all measured distances is called S.

The calculation of the distance pairs is based on the formulas given in Schneider and Eberly [Schn 02] and Eberly [Eber 11].

ii) Binning. Now, S is reduced to a sequence $\{s_b\}_1^N$ of N bins. All bins have the same bin width which is $\tilde{d} = 2\, d_E/N$. Each element of the equidistant sequence is obtained by first sorting the distance values in S based on their distances to the ellipses $E(\boldsymbol{x})$. Finally, the average over all distance pairs that belong to the b-th bin is calculated. Figure 5.1 (top right) shows the result of the binning process: the edge response orthogonal to the edge of the ellipse as a function of the bins.

iii) Differentiation. An estimate of the PSF is obtained by calculating the first derivative, $\{p_b\}_1^{N-1}$, of the edge response $\{s_b\}_1^N$, where $\{p_b\}_1^{N-1}$ is given by

$$p_b = \frac{(s_{b+1} - s_{b-1})}{2\,\tilde{d}} \,, \quad b = 1, \ldots, (N-1) \,. \tag{5.1}$$

Figure 5.1 (bottom left) shows the estimate of the PSF for the given example.

iv) Fourier transform. The MTF is obtained by calculating the Fourier transform of $\{p_b\}_1^{N-1}$. Figure 5.1 (bottom right) shows the MTF for the given example.

Judy's method allows the measurement of resolution at the location of the ellipse and along the direction $\boldsymbol{d}^\perp$. In order to reflect the highest reconstructible frequencies, the bin width $\tilde{d}$ needs to be small. Note that the described approach might be susceptible to errors especially when reconstruction artifacts close to the ellipse are present and that this approach is only suitable to evaluate the spatial resolution achieved within the neighborhood of the ellipse when the iterative reconstruction method is linear. In that case, the differentiated edge profile is equivalent to directly measuring the point response of the signal.

Throughout this dissertation, a phantom that consists only of the central low contrast ellipse within the FORBILD head phantom was used to determine the resolution. This phantom was primarily chosen to avoid reconstruction artifacts close to the selected ellipse caused by other objects inside the full FORBILD head phantom. The half axes of the ellipse are of size $a = 1.8$ and $b = 3.6$; the angle ϕ_e is zero. The ellipse itself is placed off-center at location $x_c = 0$ cm, $y_c = -3.6$ cm and was modeled with a CT number of 100 HU. Figure 5.2 (first left) shows the elliptical resolution phantom. The bin width $\tilde{d}$ was always $\tilde{d} = \Delta x/4$ and $\tilde{d} = \Delta y/4$, respectively. For display and analysis, all MTF curves were normalized so that the value at zero frequency was equal to one.

5.2.2 Computational Cost

Iterative reconstruction methods often suffer from the computational effort. The measurement of computational cost in terms of time depends strongly on the implementation. To overcome this issue, which was not a topic of this dissertation, computational cost evaluation was limited to the application of the projection matrix $\mathbf{A}$. That is, the additional effort needed to create the elements of this matrix was not included in the evaluations. Computational cost refers in this context to a standardized number of the necessary calculation steps needed to reach a given reconstruction.

The metric for computational cost was the number of iterations needed to reach a fixed resolution times a factor adjusting for differences in the number of non-zero elements in $\mathbf{A}$, which reflects the difference in computational cost associated with the application of $\mathbf{A}$. Let i be a placeholder representing a specific technique of modeling the projection matrix $\mathbf{A}$, let N_{A_i} be the number of non-zero elements of the projection matrix $\mathbf{A}_i$, and let $N_{A,\min}$ be the lowest number over all N_{A_i}, i.e., $N_{A,\min} = \min(N_{A_i})$. Then, the adjustment factor, σ_A, for each projection matrix is given by

$$\sigma_A = \frac{N_{A_i}}{N_{A,\min}} \, . \tag{5.2}$$

5.2.3   Overshoot Evaluation

Iterative reconstruction methods often yield results that include overshoot and undershoot errors at sharp edges. Often, these errors become increasingly pronounced as the number of iterations is increased. Typically, the magnitude of overshoot and undershoot errors is similar. Thus, the focus is on the evaluation of the overshoot magnitude.

The overshoot evaluation was carried out within the rectangular box that is shown in Figure 5.2 (second left). First, for each row in the given ROI the maximum reconstructed value was identified. Then, the difference between this maximum and the expected attenuation value of 50 HU was computed. Finally, the average over all the differences obtained from one row to the next was defined as the overshoot error, $\bar{\sigma}_{\text{os}}$.

5.2.4   Reconstruction Error

Mean absolute reconstruction error (bias). In practice, the reconstructed image deviates from the original image. This deviation might include both a possible offset in CT numbers and discretization errors. Let $\bar{\sigma}_b$ denote the mean absolute reconstruction error (bias) over N_b pixels that share a common attenuation value, f_0, in the ground truth image. Then, the bias is defined as the average over the absolute difference between the noise-free reconstructed values, $f_{\text{nf}}(\boldsymbol{x})$, and f_0:

$$\bar{\sigma}_b = \frac{1}{N_b} \sum_{i=1}^{N_b} |f_{\text{nf}}(\boldsymbol{x}_i) - f_0| \, . \tag{5.3}$$

The set of voxels taken into account for the bias measurement influences the value of $\bar{\sigma}_b$; for instance, voxels close to a high contrast edge might dominate the measurement. Thus, the set of pixels used throughout the dissertation was identified using a mask that is shown in Figure 5.2 (third left). To create this mask, the ground truth image was convolved with a square box function. Then, the convolved image was compared with the original phantom, and the mask was defined as the set of pixels that showed the same attenuation value of $f_0 = 50$ HU in both images. The width of the box

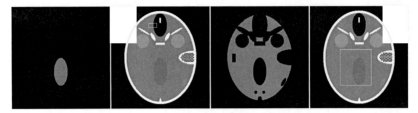

Figure 5.2: Phantom used for computation of the MTF curves (c/w=0/200 HU) *(first left)*. Region used for evaluation of the overshoot artifact; the region is the small rectangular box with white borders in the upper portion of the phantom (c/w=50/100 HU) *(second left)*. Mask used for the evaluation of the reconstruction error in noise-free reconstructions (c/w=50/100 HU) *(third left)*. Region used for evaluation of the mean standard deviation and the correlation coefficient in reconstructions from noisy data; the region is the large rectangle with white borders (c/w=50/100 HU) *(right)*.

function was equal to 9 pixels. This width was carefully selected to avoid interference with resolution effects.

Root mean square error   Another, more common method to quantify the reconstruction error is the root mean square error (RMSE), denoted as σ_{RMSE}. This measure represents the sample standard deviation of the differences and is defined as

$$\sigma_{\mathrm{RMSE}} = \sqrt{\frac{1}{N_b} \sum_{i=1}^{N_b} \left(f_{\mathrm{nf}}(\boldsymbol{x}_i) - f_0 \right)^2}. \qquad (5.4)$$

Note that very different images can have the same RMSE as shown in example in [Wang 04].

5.2.5   Noise

Due to statistical noise in the data, any reconstructed image needs to be seen as one realization of a multivariate random variable. This variable is essentially normally distributed when a large number of photons are used for the data simulation and when the effect of the reconstruction algorithm is negligible which is true for a linear reconstruction algorithm. Thus, effects due to noise could be quantified by analyzing the standard deviation in the pixel value as well as the correlations between pixels. In order to avoid significant statistical errors in the evaluation of pixel standard deviation and correlation between pixels, mean values were used as final metrics.

5.2.5.1   Mean Standard Deviation

Let $\boldsymbol{x}_i$ denote a single pixel in a given region of interest (ROI) which includes altogether N_{roi} pixels. Let f_n^j be the noise reconstructed value of the j-th noise realization minus the noise free reconstructed value, and let $\bar{f}_n(\boldsymbol{x}_i)$ be the average pixel noise over N_n noise realizations given by

$$\bar{f}_n(\boldsymbol{x}_i) = \frac{1}{N_n} \sum_{j=1}^{N_n} f_n^j(\boldsymbol{x}_i) \ . \tag{5.5}$$

The mean standard deviation, $\bar{\sigma}_n$, was determined by estimating first the pixel variance, $\widehat{\mathrm{VAR}}(\boldsymbol{x}_i)$, for each pixel within the ROI using the sample variance formula, given by

$$\widehat{\mathrm{VAR}}(\boldsymbol{x}_i) = \frac{1}{N_n - 1} \sum_{j=1}^{N_n} \left(f_n^j(\boldsymbol{x}_i) - \bar{f}_n(\boldsymbol{x}_i) \right)^2 \ . \tag{5.6}$$

Then, $\bar{\sigma}_n$ is obtained from the square root of the average over all sample variances in the given ROI,

$$\bar{\sigma}_n = \sqrt{\frac{1}{N_{\mathrm{roi}}} \sum_{i=1}^{N_{\mathrm{roi}}} \widehat{\mathrm{VAR}}(\boldsymbol{x}_i)} \ . \tag{5.7}$$

By using an ensemble of reconstructed noisy data sets, the impact of correlations is reduced in favor for the statistical accuracy. Moreover, effects due to image artifacts are removed by calculating the difference of noisy and noisefree reconstruction. The calculations were carried out for a rectangular ROI that is shown in Figure 5.2 (third left). The box was always centered on location $(x, y) = (-0.8625$ cm, -2.55 cm$)$. Depending on the pixel size, the box dimensions were 125×112 pixels for a moderate pixel size ($\Delta_{xy} = 0.0750$ cm) and 250×224 pixels for a small pixel size ($\Delta_{xy} = 0.0375$ cm).

5.2.5.2   Mean Correlation Coefficient

Let $\widehat{\mathrm{COV}}(\boldsymbol{x}_i, \boldsymbol{x}_j)$ be the sample covariance of pixels $\boldsymbol{x}_i$ and $\boldsymbol{x}_j$. By definition, the correlation coefficient, ρ, is defined as the covariance divided by the product of their individual standard deviations, σ_n,

$$\rho = \frac{\widehat{\mathrm{COV}}(\boldsymbol{x}_i, \boldsymbol{x}_j)}{\sigma_n(\boldsymbol{x}_i) \cdot \sigma_n(\boldsymbol{x}_j)} \ . \tag{5.8}$$

The mean correlation coefficient, $\bar{\rho}$, was determined by calculating the average over all correlation coefficients that were obtained in the given ROI, i.e.,

$$\bar{\rho} = \frac{1}{N_{\mathrm{roi}}} \sum_{k=1}^{N_{\mathrm{roi}}} \rho_k \ . \tag{5.9}$$

The calculation of $\bar{\rho}$ was carried out in the same ROI that was defined for the calculation of the mean standard deviation (see Fig. 5.2, third left). The evaluations were focused on correlations between pixels that are adjacent to each other, either in x, in y, or at 45 degrees. The corresponding mean correlation coefficients are denoted as follows: $\bar{\rho}_x$, $\bar{\rho}_y$, and $\bar{\rho}_{45}$.

5.3   Quantitative Evaluation: Task-Based Metrics

The goal of a task-based image quality assessment is to investigate whether or not a certain class of reconstructed images conveys diagnostic information, and to appreciate differences obtained in a quantitative evaluation based on basic metrics. Both mathematical model observers and human observers are used for such evaluations. Mathematical model observers such as the ideal (Bayesian) observer, the ideal linear observer (Hotelling observer), or model observers that predict human performance (anthropomorphic) are potential alternatives to human observers. For reasons of comparison, an observer performance measurement needs to be defined.

5.3.1   Observer Performance Measures

5.3.1.1   Area Under the Receiver Operating Characteristic Curve

The receiver operating characteristic (ROC) methodology illustrates the performance of a binary classification system. In a clinical situation, for instance, a diagnostician always has to deal with a binary classification task regarding the decision if a patient either has the suspected disease (abnormal) or not (normal). The underlying truth is called actually abnormal or actually normal when a disease is present or not, respectively. The so-called 2×2 decision matrix visualize the two-class prediction problem with the four possible outcomes: false negative (FN), false positive (FP), true negative (TN), and true positive (TP). This matrix is shown in Figure 5.3 (left).

The decision criterion is based on a continuous random variable of the two cases. Figure 5.3 (middle) shows two distributions of a normal and an abnormal case. An observer specific decision threshold parameter, t^c, divides the distributions into two parts. Patients that belong to the left side of the decision threshold are considered normal and patients on the right side are considered abnormal. For a given decision threshold the values FN, FP, TN, and TP can be computed. These values are used to calculate the true-positive fraction (TPF) and the true-negative fraction (TNF) defined as

$$\text{TPF} = \frac{\text{TP}}{\text{TP+FN}} \quad \text{and} \quad \text{TNF} = \frac{\text{TN}}{\text{TN+FP}} \, , \tag{5.10}$$

as well as the false-positive fraction (FPF),

$$\text{FPF} = \frac{\text{FP}}{\text{TN+FP}} = 1 - \text{TNF} \, . \tag{5.11}$$

The TPF, also known as sensitivity provides the probability that a detail present is correctly detected. The TNF, also known as specificity, provides the probability that a detail that is not present is correctly detected (false alarm probability). The sensitivity as a function of the FPF, TPF(FPF), is called ROC curve shown in Figure 5.3 (right). Each single point on the ROC curve correspond to a certain decision threshold t^c. Hence, by moving the decision threshold over the entire width of the two distribution, the full ROC curve is obtained [Metz 78, Metz 08, Bush 11, Oppe 11, Barr 12].

By definition, the first and last pair of points defining the ROC curve is given by (FPF,TPF)=(0,0) and (FPF,TPF)=(1,1). Hence, the ROC curve is usually displayed

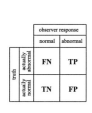 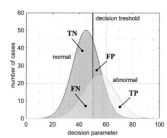 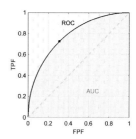

Figure 5.3: The ROC analysis is based on the decision matrix which defines the probability false negative (FN), true positive (TP), true negative (TN), and false positive (FP) *(left)*. An example for a distribution for normal and abnormal cases. For a given decision threshold (vertical line), it is possible to determine the values FN, TP, TN, and FP *(middle)*. The ROC curve illustrates the sensitivity as a function of the specificity. For a given decision threshold, a single point on the ROC curve is obtained (black dot). The entire ROC curve is obtained by moving the decision threshold over the entire width of the two distributions. The dotted line shows the ROC curve for an observer who is making a purely random decision *(right)*.

in a square of length 1. An observer making a random decision will create a diagonal line in this square (see Fig 5.3, right). All the others generate curves passing the upper diagonal of the square. Curves with a high ratio of TPF to TNF along the ROC curve are preferred.

A commonly used shorthand metric to express the overall performance of the ROC curve is the area under the ROC curve, denoted as AUC. The higher the AUC, the better is the class discrimination. The AUC may be interpreted as the probability that an observer makes a correct classification when a randomly selected pair of images from both classes are compared [Pepe 03, p. 77-78], [Barr 04, p. 823]. The AUC value range is [0.5, 1.0].

5.3.1.2   Signal-to-Noise Ratio

For further discussions, let Class 1 now contain all images where the signal is normal, and let Class 2 contain all images where the signal corresponds to a disease (abnormal signal). Let t^{ob} be a model observer statistic which is compared to the decision threshold t^c. If $t^{\mathrm{ob}} < t^c$, the image is classified as belonging to Class 1, otherwise, the observer indicates that the image belongs to Class 2 [Barr 04]. Further, let Σ_i be the covariance matrix of a given image vector of Class i, let $\bar{\Sigma} := 0.5(\Sigma_1 + \Sigma_2)$ be the average of the covariance matrices over both classes, and let μ_1 and μ_2 be the mean image under Class 1 and 2, respectively. The difference of the class means, $\Delta\mu$, is defined as $\Delta\mu = \mu_2 - \mu_1$. When the images follow Gaussian statistics and when t^{ob} corresponds to the ideal observer [Barr 04], then the signal-to-noise ratio (SNR) is defined as the difference of the class means for t^{ob} divided by the pooled standard deviation [Barr 04, p. 819]. Mathematically expressed, the square of the SNR may be computed as [Barr 04, p. 967]

$$\mathrm{SNR}^2 = \Delta\boldsymbol{\mu}^T \bar{\boldsymbol{\Sigma}}^{-1} \Delta\boldsymbol{\mu} \;. \tag{5.12}$$

The SNR is a good measure of class separability when t^{ob} is normally distributed for each class, higher values of the observer SNR indicating greater class separation [Barr 04, p. 819].

If the observer statistic, t^{ob}, is normally distributed for each class, then the AUC value, $\mathcal{A}$, may be computed from the SNR as follows

$$\mathcal{A} = \frac{1}{2}\left(1 + \mathrm{erf}\left(\frac{\mathrm{SNR}}{2}\right)\right) \;, \tag{5.13}$$

where $\mathrm{erf}(z) = 2/\sqrt{\pi} \int_0^z e^{-y^2}\mathrm{d}y$ is the conventional error function [Barr 04, p. 819].

For a finite number of images, the statistical variability becomes an issue. It is important to control and limit this variability to maximize the statistical power of IQ studies. For that reason, the known-means approach described in Wunderlich and Noo (2009 & 2013) [Wund 09, Wund 13] is employed. Wunderlich et al. showed that a large decrease in the bias and variance performance estimates could be realized by including prior knowledge of the image class means. Therefore, let p be the number of pixels centered on the signal location, let m and n be the number of measurements for Class 1 and Class 2, respectively, and let B the Euler Beta function. Then, the SNR point estimator, $\widehat{\mathrm{SNR}}$, is given by the multiplication of the SNR result obtained using the pooled sample covariance matrix, $\mathbf{S}$, with a factor, called ζ,

$$\widehat{\mathrm{SNR}} = \zeta\sqrt{\Delta\boldsymbol{\mu}^T \mathbf{S}^{-1}\Delta\boldsymbol{\mu}} \quad \text{where} \tag{5.14}$$

$$\zeta = \frac{\sqrt{\frac{2\pi}{m+n}}}{B\left(\frac{m+n-p}{2}, \frac{1}{2}\right)} \;. \tag{5.15}$$

5.3.2   Ideal Observer

The ideal observer was applied to a signal-known-exactly/background-known-exactly (SKE/BKE) binary classification task. The AUC was used as the measure of performance.

The noise in the fan-beam data was assumed to be Gaussian distributed, and the signal is chosen so that it has a negligible effect on the data statistics. Then, both image classes are characterized by the same covariance matrix. That means that the average covariance matrix is given by $\bar{\boldsymbol{\Sigma}} = \boldsymbol{\Sigma}_1 = \boldsymbol{\Sigma}_2$. Under that assumption, the AUC value $\mathcal{A}$ may be calculated by the cumulative distribution function, Φ, [Barr 04] (p. 819), [Pepe 03, p. 84], i.e., $\mathcal{A} = \Phi\left(\frac{\mathrm{SNR}}{\sqrt{2}}\right)$ with the SNR value obtained from

Equation 5.12. Taking a finite number of images into account, the AUC estimate, $\widehat{\mathcal{A}}$, is then based on the SNR point estimator of Equation 5.14, i.e.,

$$\widehat{\mathcal{A}} = \Phi \left(\frac{\widehat{\mathrm{SNR}}}{\sqrt{2}} \right) . \tag{5.16}$$

Finally, using the results in [Wund 09, Wund 12b], the standard deviation for the AUC point estimator, σ_{AUC}, can be approximated by

$$\sigma_{\mathrm{AUC}} = \Phi' \left(\frac{\widehat{\mathrm{SNR}}}{\sqrt{2}} \right) \cdot \tau \cdot \widehat{\mathrm{SNR}} \quad \text{with} \quad \tau = \sqrt{\frac{\eta}{l-1} - \frac{1}{2}} \tag{5.17}$$

where Φ' is the derivative of Φ, $l = m + n - p$, and $\eta = (l + p)\zeta^2/2$.

Both values, $\widehat{\mathcal{A}}$ and σ_{AUC}, allow the quantification and comparison of different classes of reconstructed images. Note that the ideal observer is designed to perform a specific task in an optimal way. This means that the ideal observer defines an upper performance limit by making use of all available image properties and of any prior information. Additional uncertainties in the decision making process are not introduced and the area under the ROC curve reaches a maximum [Myer 00]. The ideal observer performance is often found to be far above human performance. For that reason, either the ideal observer model may be modified (see Section 5.3.3) or human observers should be considered (see Section 5.3.4).

5.3.3 Channelized Hotelling Observer

The Channelized Hoteling Observer (CHO) was applied to a SKE/BKE binary classification task. The AUC was used as the measure of performance.

5.3.3.1 Theory

As the ideal observers, the CHO incorporates the noise information by using the covariance matrix between all images. Therefore, the CHO includes both the noise magnitude and the noise correlations which are essential for a lesion detectability task. The use of channels should reflect the constraints of the human visual system, where channelization refers to a mechanism of processing image data through finite spatial and spatial-frequency bands before forming the decision variable [Barr 04, Zhan 14]. The CHO methodology has been successfully applied in the field of medical imaging and for purposes of IQ evaluation in CT as described in LaRoque [LaRo 07].

Let $\boldsymbol{Y}$ be a column vector of size $y \times 1$ representing the noisy image realization, let C be the number of channel filters that are used to filter $\boldsymbol{Y}$, and let $\mathbf{U}$ be the impulse response of the channel filters, where $\mathbf{U}$ is of size $y \times C$. Then, the channel output vector, $\boldsymbol{v}$, is obtained by applying the impulse response to the image vector, i.e., $\boldsymbol{v} = \mathbf{U}^T \boldsymbol{Y}$.

Let $\bar{\boldsymbol{Y}}_i$ be the means of $\boldsymbol{Y}$ over Class i. Then, the means of the channel outputs is given by $\bar{\boldsymbol{v}}_i := \mathbf{U}^T \bar{\boldsymbol{Y}}_i$ and the difference of the channel output means, $\Delta \bar{\boldsymbol{v}}$, is $\Delta \bar{\boldsymbol{v}} := \bar{\boldsymbol{v}}_2 - \bar{\boldsymbol{v}}_1$. Further, let $\mathbf{K}_{Y_i}$ be the covariance matrix of $\boldsymbol{Y}$ over Class i, let $\mathbf{K}_{v_i}$

be the covariance matrix for Class i that is obtained as $\mathbf{K}_{v_i} = \mathbf{U}^T \mathbf{K}_{Y_i} \mathbf{U}$, and let $\mathbf{S}_v$ be the intraclass channel scatter matrix that is defined as

$$\mathbf{S}_v := \frac{1}{2} \left(\mathbf{K}_{v_1} + \mathbf{K}_{v_2} \right) . \tag{5.18}$$

Both $\mathbf{S}_v$ and $\Delta\bar{\boldsymbol{v}}$ are used to define the channelized Hotelling discriminant vector, $\boldsymbol{w}_{\text{Hot}}^T$, of size $C \times 1$ that is: $\boldsymbol{w}_{\text{Hot}}^T := \mathbf{S}_v^{-1} \Delta\bar{\boldsymbol{v}}$. Now, the CHO statistic, t^{ob}, is found by calculating the inner product of $\boldsymbol{v}$ with the channelized Hotelling discriminant, i.e., $t^{\text{ob}} = \boldsymbol{w}_{\text{Hot}}^T \boldsymbol{v}$.

The SNR point estimator is obtained by replacing the sample covariance matrix, $\mathbf{S}$, with Equation 5.14 with the intraclass channel scatter matrix,

$$\widehat{\text{SNR}} = \zeta \sqrt{\Delta\bar{\boldsymbol{v}}^T \mathbf{S}_v^{-1} \Delta\bar{\boldsymbol{v}}} , \tag{5.19}$$

where the definition of ζ is the same as in Equation 5.15. Based on the SNR point estimator, the AUC estimate, $\hat{\mathcal{A}}$, and the standard deviation for $\hat{\mathcal{A}}$ is obtained by applying Equations 5.16 and 5.17, respectively.

5.3.3.2 Channel Selection

In literature, there are many choices of possible channels, e.g., Abbey [Abbe 01], Eckstein [Ecks 03], and Zhang [Zhan 06]. Throughout this work, the Gabor functions were selected to model the impulse response [Ecks 03, Zhan 06]. Let (x_0, y_0) be the center of the channel, let ω_s be the spatial width of the channel, let ν_c be the center frequency of the channel, let ψ be the orientation of the channel, and let ϵ be the phase factor. Then, the Gabor functions, $G(x, y)$, can be expressed as a Gaussian modulated by a cosine function,

$$G(x, y) = \exp\left(-\frac{4 \ln 2 \left[(x - x_0)^2 + (y - y_0)^2 \right]}{\omega_s^2} \right)$$
$$\cdot \cos\left(2\pi \nu_c \left[(x - x_0) \cos\psi + (y - y_0) \sin\psi \right] + \epsilon \right) . \tag{5.20}$$

It is important that the channel center always coincides with the lesion center [Wund 08]. The choice of the channel parameters is derived from Wunderlich and Noo [Wund 08]. In total, 40 Gabor functions including four channel passbands, $\Delta\nu$, five orientations and two phases were used. That selection encompasses the following parameter set:

$$\Delta\nu = \left[\frac{1}{64}, \frac{1}{32} \right], \left[\frac{1}{32}, \frac{1}{16} \right], \left[\frac{1}{16}, \frac{1}{8} \right], \text{ and } \left[\frac{1}{8}, \frac{1}{4} \right] \text{ cycles/pixel},$$

$$\nu_c = \frac{3}{128}, \frac{3}{64}, \frac{3}{32}, \text{ and } \frac{3}{16} \text{ cycles/pixel},$$

$$\omega_s = 56.48, 28.24, 14.12, \text{ and } 7.06,$$

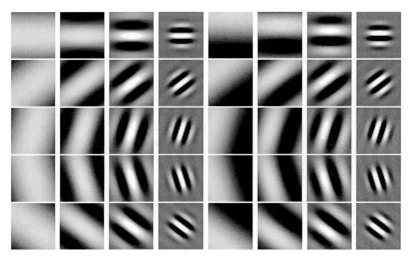

Figure 5.4: Selected Gabor channels for the CHO. Different orientations: $\psi = 0, \frac{2\pi}{5}, \frac{4\pi}{5}, \frac{6\pi}{5}, \frac{8\pi}{5}$ radians *(top to bottom)*. Frequencies $\nu_c = \frac{3}{128}, \frac{3}{64}, \frac{3}{32}, \frac{3}{16}$ cycles/pixel with $\epsilon = 0$ *(column 1 to 4)*. Frequencies $\nu_c = \frac{3}{128}, \frac{3}{64}, \frac{3}{32}, \frac{3}{16}$ cycles/pixel with $\epsilon = \frac{\pi}{2}$ *(column 5 to 8)*.

$$\psi = 0, \frac{2\pi}{5}, \frac{4\pi}{5}, \frac{6\pi}{5}, \text{ and } \frac{8\pi}{5} \text{radians},$$

$$\epsilon = 0, \text{ and } \frac{\pi}{2}.$$

The tools given in Wunderlich [Wund 14] were used to create the channels. The selected Gabor channels are shown in Figure 5.4.

5.3.3.3 Internal Noise

When human observers are asked to read the same set of images several times, for instance on different dates, the observers will make different decisions. Such variations are inevitable in human observer studies and can be attributed to internal noise [Burg 88]. To account for the variability of a human observer, an internal noise component can be included in the model observer.

There are different strategies for internal noise insertion. That includes internal noise on the decision variable or on channel level. Zhang et al. [Zhan 07] found that the best agreement between the CHO model with internal noise and the human observer performance was found when internal noise was added on channel level. Therefore, this approach was adopted in this work.

Let $\mathbf{K}_{v_i}^{\text{ext}}$ be the covariance matrix caused by external noise and let $\mathbf{K}_{v_i}^{\text{int}}$ be the covariance matrix due to internal noise. If the internal noise is directly proportional to the noise in the image (as it is indicated in Burgess and Colborne [Burg 88]), then the modified covariance matrix, $\mathbf{K}_{v_i}^{\text{int,ext}}$, can be defined as

$$\mathbf{K}_{v_i}^{\text{int,ext}} = \mathbf{K}_{v_i}^{\text{ext}} + \mathbf{K}_{v_i}^{\text{int}} \quad \text{with} \tag{5.21}$$

$$\mathbf{K}_{v_i}^{\text{int}} = \xi \ \text{diag}(\mathbf{K}_{v_i}^{\text{ext}}), \tag{5.22}$$

where ξ is a free parameter that specifies the amount of internal noise addition. In general, the external noise matrix is identical to $\mathbf{K}_{v_i}$. The AUC estimate is obtained by first replacing $\mathbf{K}_{v_i}$ with $\mathbf{K}_{v_i}^{\text{int,ext}}$ in Equation 5.18 and then following the approach as described in Section 5.3.3.1.

5.3.4   Human Observer

Human observers assessed IQ in a two alternative forced choice (2-AFC) experiment. In a 2-AFC experiment, the observer is consecutively shown a pair of images where only one of the images contains the lesion. The lesion was always circular with a fixed diameter, random contrast, and random location. To make the observer familiar with the visual lesion properties, each 2-AFC experiment incorporated a training session followed by testing session. In the sessions, the observer was asked to indicate the lesion location over a graphic user interface that was written in Matlab (Mathworks Inc., Natick, MA) and provided by Wunderlich (Food and Drug Administration, FDA) and Noo (University of Utah). Figure 5.5 shows the design of the user interface for (top left) the training and (bottom right) the testing session. The selected lesion location is indicated by a red cross. During the training session, the observer received a feedback of the lesion location indicated by a green box. This feedback loop was omitted in the testing session. Note that the gray scale window was fixed for all observers. To avoid additional uncertainties, all observers read the images in the same dimmed dark room and on the same DICOM calibrated screen.

The 2-AFC experiment aim to estimate the probability of making a correct decision. A correct localization decision occurs if and only if the location of the lesion is detected correctly. This probability is used as the figure of merit (FOM) and corresponds to the area under the localization ROC (LROC) curve. The probability is typically estimated as a proportion of correct detected lesion locations and the total number of image pairs that was shown to the reader. This proportion can serve as an unbiased estimate of the probability of making a correct decision. That probability highly depends on the observer, the selected cases, and the total number of image pairs shown. The statistical variability can be reduced with an increasing number of image pairs as the proportion comes closer to the desired probability of correct decision [Noo 13].

Swensson proposed this simple formal model of describing the performance of observers who are required to find and locate possible targets on images [Swen 96]. The LROC curve is obtained by plotting the true localized positive fraction as a function over the false localized positive fraction. Due to the forced choice experiment, the end point of the LROC curve is not necessary $(1, 1)$, and the area under the LROC curve (AUC_L) is bounded in the interval $[0, 1]$. Figure 5.5 gives an example for an

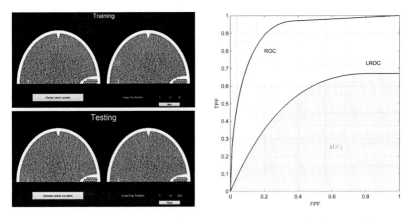

Figure 5.5: For each image pair, the lesion is present in one and only one of the two images. The observer was asked to select one lesion location within one of the two images *(left)*. User interface for training and testing *(top left, top bottom)*. During the training session the observer decided that the lesion was in the left image (red cross), whereas the lesion was located in the right image (green box). The green box is not present for the testing session. Gray scale: $[-150; 250]$ HU. The figure shows the differences between a ROC and a LROC curve and highlights the area under the LROC curve, denoted in this figure as AUC_L *(right)*.

ROC curve versus a LROC curve. A more detailed and mathematical description of the ROC versus the LROC analysis is described in Popescu [Pope 07].

5.4 Basic Metrics versus Task-based Metrics

There is always a discussion whether basic metrics are sufficient for an assessment of image quality. Irrespective of that discussion, all evaluations should always be carried out in images that show the same resolution. However, both basic metrics and task-based metrics have weaknesses and strengths that will be discussed briefly below.

The main advantage of basic metrics is the ease of use and transferability to any reconstructed image. Such metrics are completely objective. However, a complication of basic metrics is, for example, that they are very sensitive to small changes in scale or orientation. This means that a change of the ROI where the evaluation is carried out may, at worst, change the whole conclusion. For example, by using an ROI that includes edges in the phantom, the bias evaluation may be influenced by possible over- and undershoots that occur in iterative reconstruction. Therefore, the estimate of the bias might be distorted. Further, the visual impression of an image to the reader may be different between two images; however, the basic metric measurements do not necessarily indicate that difference.

Task-based metrics may help to predict human observer performance more accurately. Such assessments are clinically relevant but they are extremely time consuming and difficult to control. In order to improve the statistical variability many observers and images are required. An ideal observer is often seen as a best-case scenario. A human most likely will never reach the performance of an ideal observer. Therefore, some people may ask to include additional channels or internal noise to lower the measurement of performance. Such properties may reflect a performance that is close to a human. However, it is difficult to say how close the final performance measure is compared to human. Human observer studies prevent discussions whether one or another mathematical model would have been the best choice. However, that implies the ability to find a group of skilled people who are motivated enough to go through a large number of images. Not only for that reason, but also because computerized model observers offer an attractive alternative for the evaluation of lesion detectability, mathematical models are often preferred.

CHAPTER 6

Impact of Discrete Image Representation Techniques on Image Quality

This chapter describes the impact of discrete image representation techniques on image quality. First, both the motivation and limitations of this study are described in Section 6.1. Next, in Section 6.2, the mathematical formulation of various image representation techniques that were investigated are explained. This section is followed by Section 6.3, which gives a description of the experimental comparison conditions, specifically of the reconstruction method, the geometrical settings, and the phantom. Section 6.4 shows the results obtained in a preliminary study including all forward projection models presented that primarily focused on a qualitative assessment of image quality. Based on the results of the preliminary study, an extensive evaluation of IQ was set up based on a small selection of three linear forward projection models, specifically Joseph's method, the distance-driven method, and the bilinear method. Section 6.5 and 6.6 present the results of the quantitative evaluation using basic metrics and task-based metrics, respectively. Finally, a discussion and conclusion is given in Section 6.7

Parts of this work have already been published in Schmitt et al. [Schm 12b, Schm 13, Schm 14a] and Hahn et al. [Hahn 15a, Hahn 16].

6.1   Motivation and Limitations

Non-linear IR methods have become very popular in the CT community. As discussed in Section 4.4, these reconstruction methods are often linked with an incredibly large number of degrees of freedom. In order to gain the most out of non-linear IR methods, multi-directional investigations are required. This also includes a better understanding of how various parameters impact IQ. As shown for instance in Matej and Lewitt (and references therein) [Mate 96], the properties of the reconstructed image can significantly vary with the forward projection model and with the parameters defining it. Conceptually, the choice of the forward projection model is a major step in the design of an IR algorithm, particularly because the decision being made at this level affects both bias and noise properties of the reconstruction. In addition, the selection of additional parameters that might appear in the cost function or the regularization also influence IQ. However, optimizing a variety of parameters is complicated and accompanied with the need to perform several reconstructions to account for different noise realizations and variations in geometry, which is essential for meaningful observations. To mitigate this difficulty the focus of the study was on 2D rather than on 3D reconstructions. The hope is that the acquired knowledge extends at least partially to 3D.

Some preliminary observations shown in Schmitt et al. [Schm 12b] and Section 6.4 motivated us to set up an extensive IQ assessment study based on linear interpolation models. Linear interpolation models are typically seen as a suitable tradeoff between discretization errors and computational effort for image reconstruction in CT. An extension of this study including other forward projection models was not possible due to the high computational effort of generating a variety of different scanning geometries along with a fair number of repeated scans in each geometry. The same applies to an extension to 3D.

6.2   Discrete Image Representation Techniques

Different techniques of modeling the forward projection matrix exist. Commonly used techniques to represent the image with a finite number of unknowns are the sampling approach, the strip integral approach and the basis function approach. All approaches are described below in two dimensions. In most cases the extension to three dimensions is straightforward but more difficult to illustrate.

6.2.1   Sampling Approach

In the sampling approach, as the name already suggests, the image is represented by its values at a fixed number of samples. These locations are typically equidistantly distributed in the direction of Cartesian coordinates [Buzu 08]. A very popular sampling approach is the method of Joseph.

The Joseph's method was suggested by Joseph in 1982 [Jose 82]. In principle, this method models each measurement as a straight line integral. The line integral is approximated as a summation of weighted pixel contributions along the line. The slope of the line as defined by the projection angle θ (Eq. 3.13) determines whether

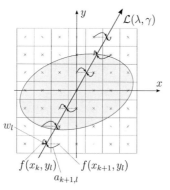

Figure 6.1: Schematic illustration of Joseph's method for a line $\mathcal{L}$ that is more parallel to the y axis. The measurement is approximated as a summation of weighted pixel contributions, i.e., $g(\lambda, \gamma) \simeq \sum_{k,l} a_{k(l),l} f(x_k, y_l)$. Pixels that do not contribute are weighted with $a_{k,l} = 0$. The parameter w_l defines the distance in x between the intersection of the line $\mathcal{L}$ with the line of coordinate y_l and the pixel location $f(x_k, y_l)$.

the summation is in x or in y direction. In detail, when $|\cos\theta| < 1/\sqrt{2}$, the line is more parallel to the x-axis than to the y-axis and the summation is in x; otherwise, it is in y. The interpolation direction is perpendicular to the summation direction. This means, when the summation is in x (alternatively y), the interpolation is in y (respectively x). The weights are determined using linear interpolation.

The concept of Joseph's method is illustrated in Figure 6.1 for a line that is more parallel to the y axis. Let k and l be the indices of the Cartesian coordinates x and y, respectively. In the example given by Figure 6.1, for each l, there is an intersection between $L(\lambda, \gamma)$ and the line of coordinate y_l, and there is an index $k(l)$ such that this intersection is between $x_{k(l)}$ and $x_{k(l)+1}$ in x. Further, let w_l be the distance in x between the intersection and $x_{k(l)}$. Then, the underlying formula for Joseph's method for a line that is more parallel to the y axis can be expressed as

$$g_f^J(\lambda, \gamma) \simeq \frac{\Delta y}{|\cos\theta|} \sum_l^{N_y} \left\{ \frac{\Delta x - w_l}{\Delta x} f(x_{k(l)}, y_l) + \frac{w_l}{\Delta x} f(x_{k(l)+1}, y_l) \right\} . \qquad (6.1)$$

Whenever $x_{k(l)}$ and $x_{k(l)+1}$ are outside the range of available samples the addend is zero. The scaling factor in front of the summation is the Jacobian that accounts for the summation being performed in y instead of along the line.

For the case that the line is more parallel to the x axis, let w_k be the distance in y between the intersection of $\mathcal{L}$ and $y_{l(k)}$. Then, the formula is given by

$$g_f^J(\lambda, \gamma) \simeq \frac{\Delta x}{|\sin\theta|} \sum_k^{N_x} \left\{ \frac{\Delta y - w_k}{\Delta y} f(x_k, y_{l(k)}) + \frac{w_k}{\Delta y} f(x_k, y_{l(k)+1}) \right\} , \qquad (6.2)$$

which is basically a modified version of Equation 6.1, but with a summation in x and with an dependence on the index k. The same constraints apply, i.e., whenever $y_{l(k)}$ and $y_{l(k)+1}$ are outside the range of available samples the corresponding addends are zero.

The definition of Joseph's method implies that the direction of summation may change within any given fan-beam projection. Alternatively, the direction of summation for all measurements within one view could be fixed according to the polar angle λ. Intuitively, a better accuracy may be expected when the direction of summation is individually selected for each ray, particularly for geometries with a high magnification factor. For that reason, this approach was used throughout this dissertation. In literature, there is no consensus about that issue because Joseph's method was truly described for parallel-beam data.

6.2.2   Strip Integral Approach

The distance-driven method has become more and more popular over the last years and is a strip integral approach. The distance-driven method was suggested by De Man and Basu in 2002 [De M 02] and is based on the same concept as the method of Joseph. In addition, the distance-driven method accounts for the finite detector pixel size. As in Joseph's method, each measurement is approximated using a summation in x or y of weighted pixel contributions along the line. The direction of summation is defined by the polar angle and is the same for all rays within a same view. In detail, if $\sin \lambda < 1/\sqrt{2}$, the summation is in x, otherwise in y. When the summation is in x (alternatively y), the contributing samples in the summation are defined by the projected length of the detector pixel on each column (respectively row) of pixels. The weight of each contribution is determined using a nearest-neighbor interpolation model.

The concept of the distance-driven method for a case where the summation is in y is illustrated in Figure 6.2. To create the measurement associated with a given detector pixel, two lines that connect the source position with the edges of the detector are required. In the given example of Figure 6.2, for each l, these two lines intersect the line of coordinate y_l at two locations in x, called $q_1(l)$ and $q_2(l)$, with $q_1(l) < q_2(l)$. These two locations define the projected length of the detector pixel over which integration is to be carried out. This integration is performed in x using a nearest-neighbor interpolation model for representation of $f(x_k, y_l)$. Let $d_{k,l}$ be the length of the overlap between the interval $[q_1(l), q_2(l)]$ and $[x_k - \Delta x/2, x_k + \Delta x/2]$, where the latter interval is the length of the image pixel at location x_k. Then, the formula of the distance-driven method for a line that is more parallel to the y axis can be expressed as

$$g_f^D(\lambda, u) \simeq \frac{\Delta y}{|\cos \theta|} \sum_l^{N_y} \left\{ \sum_k^{N_x} \frac{d_{k,l}}{q_2(l) - q_1(l)} f(x_k, y_l) \right\}. \qquad (6.3)$$

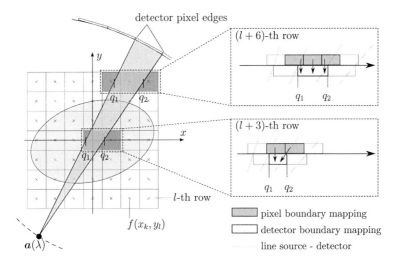

Figure 6.2: Schematic illustration of the distance-driven method for a line that is more parallel to the y axis. The measurement is approximated as a summation of weighted pixel contributions. The pixels that do not contribute are weighted with zero. The contributing pixels are defined by the projected length of the detector pixel on each row, indicated as $[q_1, q_2]$. The projected length $[q_1, q_2]$ and the number of contributing samples along the row that are involved in the summation vary with the pixel width as illustrated on the right side.

For the case that the summation is carried out in x direction, the formula is given by

$$g_f^D(\lambda, u) \simeq \frac{\Delta x}{|\sin\theta|} \sum_k^{N_x} \left\{ \sum_l^{N_y} \frac{d_{k,l}}{q_2(k) - q_1(k)} f(x_k, y_l) \right\}. \tag{6.4}$$

If no overlap between the intervals exists, the length $d_{k,l}$ is equal to zero. The scaling factors in front of the summations in the last two equations are the same as in Joseph's method and they have the same meaning. By definition, the inner summation over k in Equation 6.3 and over l in Equation 6.4 involve only a few terms. The exact number of that terms depends on the pixel size, as well as on l, λ, u and the detector pixel size Δu. The drawings on the right side in Figure 6.2 illustrate how the number of terms involved in the summation may vary within one measurement.

6.2.3 Basis Function Approach

In the basis function approach, the image is represented by a finite linear combination of specific functions that are often selected as scaled and translated versions of a single function, called the mother function. Since the basis function approach yields a continuous model for the image, the estimate of the CT measurements is usually straightforward.

6.2.3.1   General Concept

Let $f_a(x,y)$ denote the estimate of $f(x,y)$ and let $b(x,y)$ denote the mother function. Then, the estimate of the image can be written in the following form:

$$f_a(x,y) = \sum_{k}^{N_x} \sum_{l}^{N_y} f(x_k, y_l) \cdot b\left(\frac{x - x_k}{\Delta x}, \frac{y - y_l}{\Delta y} \right). \tag{6.5}$$

Using the above expression for f_a together with the parallel-beam coordinates θ and s, the estimate of a measurement is simply approximated by the Radon transform of $f_a(x,y)$, denoted as $g_p^a(\theta, s)$. Let $r(\theta, s)$ be the Radon transform of $b(x,y)$, then due to the linearity of the Radon transform the general expression for any kind of basis function is

$$g_p^a(\theta, s) = \sum_{k}^{N_x} \sum_{l}^{N_y} f(x_k, y_l) \cdot r\left(\theta, \frac{s - x_k \cos\theta - y_l \sin\theta}{\Delta_{xy}} \right). \tag{6.6}$$

Unlike to Joseph's method and the distance-driven method, the basis function concept does not necessarily require a selection of a preferred Cartesian direction for the calculation of the measurements. Generally, the basis function approach is more appealing than the sampling approach due to the continuous model. However, in terms of computational efficiency, the sampling approach might be more attractive.

6.2.3.2   B-splines

The B-splines [Horb 02] have become one of the most popular mother functions. The B-splines are simple piecewise polynomial functions defined by a single parameter: the degree, n, of the polynomial. Let β^0 be the normalized box function:

$$\beta^0(x) = \begin{cases} 1, & -\frac{1}{2} \leq x \leq \frac{1}{2} \\ 0, & \text{otherwise} . \end{cases}$$

Then, the centered B-spline, β^n, is the $(n+1)$-th convolution of the normalized box function with itself, i.e.,

$$\beta^n(x) = \beta^0 * \beta^{n-1}(x)$$
$$= \underbrace{\beta^0 * \cdots * \beta^0}_{(n+1)\,\text{factors}} . \tag{6.7}$$

In order to rewrite the equation for β^n in a more convenient form as it was found in [Horb 02], let the one-sided power function, x_+^n, be defined as

$$
x_+^n = \begin{cases} x^n, & x \geq 0 \text{ and } n > 0 \\ 1, & x \geq 0 \text{ and } n = 0 \, , \\ 0, & \text{otherwise} \end{cases} \tag{6.8}
$$

and, let c_{n+1}^k define the number of different ways of distributing k successively in a sequence of $n + 1$ trials, i.e.,

$$
c_{n+1}^k = \binom{n + 1}{k} \, , \tag{6.9}
$$

and further let $d_{n,k}$ and $\alpha_{n,k}$ be

$$
d_{n,k} = k - \frac{(n + 1)}{2} \, , \tag{6.10}
$$

$$
\alpha_{n,k} = (-1)^k \, c_{n+1}^k \, . \tag{6.11}
$$

Then, using the definitions above, Equation 6.7 can also be expressed as

$$
\beta_h(x) = \frac{1}{n!} \sum_{k=0}^{n+1} \alpha_{n,k} \, [x - d_{n,k}]_+^n \, . \tag{6.12}
$$

The B-splines mother function, b^{n_1,n_2}, in two dimensions is then defined as the tensor product of two B-splines of degree n_i with $i = 1, 2$, i.e.,

$$
b^{n_1,n_2}(x, y) = \beta^{n_1}(x) \, \beta^{n_2}(y) \, . \tag{6.13}
$$

Figure 6.3 illustrates the continuous B-spline basis function for $n := n_1 = n_2$. The higher the order n, the smoother becomes the function b^{n_1,n_2} and the more resolution is lost. Note that using the B-spline of order $n = 0$ for image representation is equivalent to adopting the approach of Siddon [Sidd 85]. The B-splines of order $n = 1$ corresponds to the concept of the bilinear method (see below, p. 60).

Horbelt et al. [Horb 02] showed that the Radon transform, r^{n_1,n_2}, of the two-dimensional basis function b^{n_1,n_2} is a convolution of two B-splines:

$$
r^{n_1,n_2}(\theta, s) = \left(\beta_{|\cos\theta|}^{n_1} * \beta_{|\sin\theta|}^{n_2} \right)(s) \, . \tag{6.14}
$$

Furthermore, using the convolution properties of the B-splines and the definitions of $d_{n,k}$ (Eq. 6.10) and $\alpha_{n,k}$ (Eq. 6.11), they also showed that the Radon transform can be formulated as

$$
r^{n_1,n_2}(\theta, s) = \sum_{p=0}^{n_1+1} \sum_{q=0}^{n_2+1} \alpha_{n_1,p} \, \alpha_{n_2,q} \, \frac{\left[s - d_{n_1,p} \, |\cos\theta| - d_{n_2,q} \, |\sin\theta| \right]_+^{n_1+n_2+1}}{(n_1 + n_2 + 1)! \, |\cos\theta|^{n_1+1} \, |\sin\theta|^{n_2+1}} \, , \tag{6.15}
$$

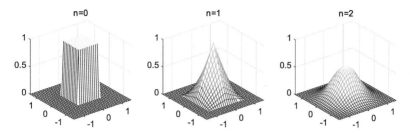

Figure 6.3: Illustration of the B-spline mother function $b^{n_1,n_2}(x,y)$ with $n := n_1 = n_2$. $n = 0$ *(left)*, $n = 1$ *(middle)*, and $n = 2$ *(right)*. The sampling in x and y was assumed to be the same for all plots.

with the following limit values:

$$r^{n_1,n_2}(\theta = 0,\, s) = \beta^{n_1}(s)\,, \tag{6.16}$$

$$r^{n_1,n_2}\left(\theta = \frac{\pi}{2},\, s\right) = \beta^{n_2}(s)\,. \tag{6.17}$$

The limits can be easily established from Equation 6.14.

The computation of r^{n_1,n_2} is expensive, especially for large n. This cost could be largely decreased using interpolation with a look-up table. However, such a table was not used throughout this dissertation, as the primary goal was to assess the best possible performance.

Bilinear Method. The bilinear method was suggested by Kak and Slanley[Kak 01, Section 7.4], but was, to the best knowledge of the author, never tested without approximation. The concept behind the bilinear method is to use bilinear interpolation to estimate the value of $f(x,y)$ at all points along the line defining the X-ray measurements, and then integrate (sum) these values together.

Using the B-splines basis function approach, the bilinear method is received for $n = 1$. For equal n and h, the expression of the normalized box function $\beta^n(x)$ in Equation 6.12 simplifies to

$$\beta^1(x) = \begin{cases} 1 - |x|, & |x| < 1 \\ 0, & \text{otherwise}, \end{cases} \tag{6.18}$$

and, therefore, the mother function is $b^{1,1}(x,y) = \beta^1(x)\,\beta^1(y)$. Then, the Radon transform of the bilinear method is

$$r^{1,1}(\theta, s) = \sum_{p=0}^{2}\sum_{q=0}^{2}(-1)^{p+q}\binom{2}{p}\binom{2}{q}\frac{[s-(p-1)\,|\cos\theta| - (q-1)\,|\sin\theta|]_{+}^{3}}{3!\,|\cos\theta|^{2}\,|\sin\theta|^{2}}\,. \tag{6.19}$$

Novel formula. Although it is analytically attractive, Equation 6.15 is unfortunately numerically unstable for values of θ that are close to (but different from) zero and $\pi/2$. In addition, the instability quickly increases with increasing n_1 and n_2. The core of the problem is that the summation in Equation 6.15 must factor out $|\sin\theta|^{n+1}$ when θ is close to zero, and $|\cos\theta|^{m+1}$ when θ is close to $\pi/2$, whereas these quantities can be beyond machine precision at some point. For this purpose, we introduce the following novel formula that is numerically stable.

First, let

$$\xi_n(x) = \begin{cases} 1, & x < d_{n,n+1} \\ 0, & \text{otherwise .} \end{cases} \tag{6.20}$$

Then, Equation 6.12 for the normalized box function $\beta^n(x)$ can be rewritten in the following form

$$\beta^n(x) = \frac{1}{n!} \sum_{k=0}^{n} \alpha_{n,k} \left[x - d_{n,k}\right]_+^n \xi_n(x) . \tag{6.21}$$

Using the last equation together with Equation 6.13, the expression for the mother function is

$$b^{n_1,n_2}(x,y) = \beta^{n_1}(x)\,\beta^{n_2}(y)$$
$$= \frac{1}{n_1!}\frac{1}{n_2!} \sum_{p=0}^{n_1} \sum_{q=0}^{n_2} \alpha_{n_1,p}\,\alpha_{n_2,q} \left[x - d_{n_1,p}\right]_+^{n_1} \left[y - d_{n_2,q}\right]_+^{n_2} \xi_{n_1}(x)\,\xi_{n_2}(y) . \tag{6.22}$$

The inner terms of Equation 6.22 that depend on x, y, p and q can be grouped together in a new function, called $b_{p,q}^{n_1,n_2}(x,y)$, i.e.,

$$b_{p,q}^{n_1,n_2}(x,y) = \left[x - d_{n_1,p}\right]_+^{n_1} \left[y - d_{n_2,q}\right]_+^{n_2} \xi_{n_1}(x)\,\xi_{n_2}(y) \tag{6.23}$$
$$= \left[x\right]_+^{n_1} \left[y\right]_+^{n_2} \xi_{n_1}(x + d_{n_1,p})\,\xi_{n_2}(y + d_{n_2,q}) , \tag{6.24}$$

where Equation 6.24 was obtained from Equation 6.23 by substituting the arguments of the one-sided power function.

Next, let $r_{p,q}^{n_1,n_2}$ be the Radon transform of $b_{p,q}^{n_1,n_2}$. Then, the Radon transform of Equation 6.22 is given by

$$r^{n_1,n_2}(\theta,s) = \left(\mathcal{R}b^{n_1,n_2}\right)(\theta, s - d_{n_1,p}\cos\theta - d_{n_2,q}\sin\theta)$$
$$= \frac{1}{n_1!}\frac{1}{n_2!} \sum_{p=0}^{n_1} \sum_{q=0}^{n_2} \alpha_{n_1,p}\,\alpha_{n_2,q} \left(\mathcal{R}b_{p,q}^{n_1,n_2}\right)(\theta, s - d_{n_1,p}\cos\theta - d_{n_2,q}\sin\theta)$$
$$= \frac{1}{n_1!}\frac{1}{n_2!} \sum_{p=0}^{n_1} \sum_{q=0}^{n_2} \alpha_{n_1,p}\,\alpha_{n_2,q}\, r_{p,q}^{n_1,n_2}(\theta, s - d_{n_1,p}\cos\theta - d_{n_2,q}\sin\theta) . \tag{6.25}$$

Now, the focus is on the calculation of $\mathcal{R}b_{p,q}^{n_1,n_2}$. First, the general definition of the Radon transform of $b_{p,q}^{n_1,n_2}(x,y)$ is

$$r_{p,q}^{n_1,n_2}(\theta,s) = \int_{-\infty}^{\infty} b_{p,q}^{n_1,n_2}(s\cos\theta - t\sin\theta, s\sin\theta + t\cos\theta)\, dt , \tag{6.26}$$

which can be rewritten using Joseph's expression, i.e.,

$$r_{p,q}^{n_1,n_2}(\theta,s) = \begin{cases} \frac{1}{|\sin\theta|} \int_{-\infty}^{\infty} b_{p,q}^{n_1,n_2}\left(x, \frac{s-x\cos\theta}{\sin\theta}\right) dx, & \text{if } |\sin\theta| > |\cos\theta| \\ \frac{1}{|\cos\theta|} \int_{-\infty}^{\infty} b_{p,q}^{n_1,n_2}\left(\frac{s-y\sin\theta}{\cos\theta}, y\right) dy, & \text{if } |\sin\theta| < |\cos\theta| . \end{cases} \tag{6.27}$$

Note that $r_{p,q}^{n_1,n_2}(\theta,s)$ is fully defined from its values over $(\theta,s) \in [0,\frac{\pi}{2}] \times [0,\infty)$. Since $b^{n_1,n_2}(x,y) = b^{n_1,n_2}(-x,y) = b^{n_1,n_2}(-x,-y)$, the following relations are obtained: i) $r(\theta,s) = r(\theta,-s)$ for $s < 0$, and ii) $r(\theta,s) = r(\pi-\theta,s)$ for $\theta \in (\frac{\pi}{2},\pi)$. For other values of θ and s, the classical property of parity apply $r(\theta+\pi,-s) = r(\theta,s)$ (see also Sec. 3.2). If $n_1 = n_2$ the Radon transform is fully defined in $[0,\frac{\pi}{4}] \times [0,\infty)$, since $b^{n_1,n_2}(x,y) = b^{n_1,n_2}(y,x)$.

Second, from Equation 6.24 it becomes apparent that $b_{p,q}^{n_1,n_2}(x,y)$ is compactly supported over the region

$$(x,y) \in [0, \underbrace{d_{n_1,n_1+1} - d_{n_1,p}}_{u}] \times [0, \underbrace{d_{n_2,n_2+1} - d_{n_2,q}}_{v}], \tag{6.28}$$

with $v = u \tan\phi$.

Then, using the definition of $b_{p,q}^{n_1,n_2}(x,y)$ (Eq. 6.24) and the piecewise definition of the Radon transform (Eq. 6.27), the definition of $r_{p,q}^{n_1,n_2}$ with $(\theta,s) \in [0,\frac{\pi}{2}] \times [0,\infty)$ is given as follows:

Case I: $0 \le \theta \le \phi,\ 0 \le s \le u\cos\theta + v\sin\theta$

$$r_{p,q}^{n_1,n_2}(\theta,s) = \frac{1}{\cos\theta} \int_{y_{\min}}^{y_{\max}} \frac{(s-y\sin\theta)^m}{(\cos\theta)^m} y^n\, dy \tag{6.29}$$

$$= \frac{(-1)^m}{(\cos\theta)^{m+1}} \int_{y_{\min}}^{y_{\max}} \sum_{k=0}^{m} c_m^k\, y^{n+k}(\sin\theta)^k\, (-1)^{m-k}\, s^{m-k}\, dy \tag{6.30}$$

$$= \frac{1}{(\cos\theta)^{m+1}} \sum_{k=0}^{m} c_m^k\, (-1)^k\, (\sin\theta)^k\, s^{m-k} \frac{(y_{\max}^{n+k+1} - y_{\min}^{n+k+1})}{n+k+1}, \tag{6.31}$$

with

$$y_{\min} = \begin{cases} 0, & 0 \le s \le u\cos\theta \\ \frac{s-u\cos\theta}{\sin\theta}, & u\cos\theta \le s \le u\cos\theta + v\sin\theta, \end{cases} \tag{6.32}$$

and

$$y_{\max} = \begin{cases} v, & v\sin\theta \le s \le u\cos\theta + v\sin\theta \\ \frac{s}{\sin\theta}, & 0 \le s \le v\sin\theta. \end{cases} \tag{6.33}$$

Case II: $\phi \leq \theta \leq \pi/2, \quad 0 \leq s \leq u\cos\theta + v\sin\theta$

$$r_{p,q}^{n_1,n_2}(\theta, s) = \frac{1}{\sin\theta} \int_{x_{\min}}^{x_{\max}} \frac{(s - x\cos\theta)^n}{(\sin\theta)^n} x^m \, dx \tag{6.34}$$

$$= \frac{1}{(\sin\theta)^{n+1}} \int_{x_{\min}}^{x_{\max}} \sum_{k=0}^{n} c_n^k \, x^{m+k} (\cos\theta)^k \, (-1)^k \, s^{m-k} \, dx \tag{6.35}$$

$$= \frac{1}{(\sin\theta)^{n+1}} \sum_{k=0}^{n} c_n^k \, (-1)^k \, (\cos\theta)^k \, s^{n-k} \frac{(x_{\max}^{m+k+1} - x_{\min}^{m+k+1})}{m+k+1}, \tag{6.36}$$

with

$$x_{\min} = \begin{cases} 0, & 0 \leq s \leq v\sin\theta \\ \frac{s - v\sin\theta}{\cos\theta}, & v\sin\theta \leq s \leq u\cos\theta + v\sin\theta, \end{cases} \tag{6.37}$$

and

$$x_{\max} = \begin{cases} u, & u\cos\theta \leq s \leq u\cos\theta + v\sin\theta \\ \frac{s}{\cos\theta}, & 0 \leq s \leq u\cos\theta. \end{cases} \tag{6.38}$$

The novel formula was implemented and tested in various cases. Since it was numerically more stable than Equation 6.15, the proposed approach was used throughout the dissertation unless otherwise stated.

6.2.3.3 Blobs

The blobs are a well-known example of rotationally symmetric basis functions [Lewi 90, Lewi 92]. Note, that the blobs are also known as generalized Kaiser-Bessel functions. Let r be the radial distance from the blob center, let a be the radius of the basis function, let α be a parameter controlling the blob shape, and let I_m be the modified Bessel function of order m. Then, the blob function, $\Gamma_{m,a,\alpha}(r)$, is given by the following one-dimensional expression

$$\Gamma_{m,a,\alpha}(r) = \begin{cases} \left(1 - \frac{r^2}{a^2}\right)^{\frac{m}{2}} I_m\left(\alpha\sqrt{1 - \frac{r^2}{a^2}}\right) I_m^{-1}(\alpha), & 0 \leq r \leq a \\ 0 & \text{otherwise}. \end{cases} \tag{6.39}$$

In the context of a basis function approach, the 2D mother function $b(x, y)$ is chosen as $\Gamma_{m,a,\alpha}(r)$ with $r = \sqrt{x^2 + y^2}$. Following the recommendations in [Mate 96], the parameters that were used throughout the dissertation were: $m = 2$, $a = 2$ and $\alpha = 10.4$. Figure 6.4 illustrates the continuous blob basis function using the previously mentioned parameters.

The Radon transform of the blob basis functions, called $r_{\text{blb}}(\theta, s)$, is independent of θ and is zero for $|s| > a$. Given the above definitions, $r_{\text{blb}}(\theta, s)$ can be expressed as follows

$$r_{\text{blb}}(\theta, s) = \begin{cases} \left(\frac{2a^2\pi}{\alpha}\right)^{\frac{1}{2}} \left(1 - \frac{s^2}{a^2}\right)^{\frac{m}{2} + \frac{1}{4}} I_{m+\frac{1}{2}}\left(\alpha\sqrt{1 - \frac{s^2}{a^2}}\right) I_m^{-1}(\alpha), & |s| < a \\ 0, & \text{otherwise}. \end{cases} \tag{6.40}$$

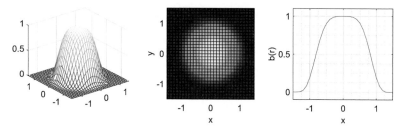

Figure 6.4: Illustration of the blob basis function $b(x, y)$ with $m = 2$, $a = 2$ and $\alpha = 10.4$ in a lateral view *(left)* and a top view *(middle)*. Radial profile of the blob basis function *(right)*. The sampling in x and y was assumed to be the same.

As for the B-spline approach, the computation of r_{blb} is expensive. However, a look-up table was not used throughout the dissertation, as the primary goal was to assess the best possible performance.

6.2.4 Comparative Discussion

A good discrete image representation technique tries to i) optimize the representation of a constant function; ii) allow for cost-effectively implementation of the projection matrix; and, iii) suppress aliasing artifacts due to high frequency components. These requirements are affected by the manner in how each approach represent the image, which can significantly impact IQ and computational cost in IR. The following comparative discussion focuses on a comparison of the differences between Joseph's method, the distance-driven method, and the bilinear method. That selection already shows fundamental differences, even though all three methods can be classified as linear interpolation models.

A first fundamental difference is about the utilization of pixel values, shown in Figure 6.5 for Joseph's method and the distance-driven method. In that example, the pixel size is small relative to the separation between the rays. In Joseph's method, such a situation yields imbalances in the sense that some pixel values appear underutilized in comparison with others. In detail, some pixels do not contribute in a particular view, as illustrated on the right side of the figure where no interpolation between pixels b and c is carried out. For a much smaller pixel sampling or for a higher magnification, it could even happen that some pixels would play no role in modeling the measurements in this view. This observation may be seen as an undesirable feature of Joseph's method. Given its definition, the bilinear method exhibits the same feature, albeit not with the same importance for a given pixel size, since the size of the pixel in both x and y matters for the bilinear method. The distance-driven method is not similarly affected.

A second fundamental difference is in the number of pixels involved in the interpolation procedure in the estimate of the measurements. In the distance-driven method, this number can vary a lot, as shown in Figure 6.5 (right), for a high magnification geometry with λ close to $\pi/4$. In that example, the interpolation length

a) Joseph's method   b) distance-driven method

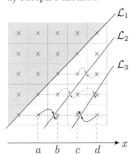
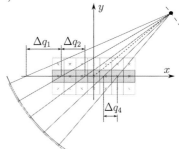

Figure 6.5: Geometrical features affecting the forward projection model. In Joseph's method pixels be under-utilized; e.g., in this view, no interpolation is required between pixels b and c *(left)*. In the distance-driven method the interpolation lengths can become quite large; e.g., in this view, Δq_1 is much larger for the negative values of u than for the positive ones *(right)*.

Δq_1 can become quite large, meaning that many pixels contribute to the estimation of the line integral. Such imbalances may negatively impact the computational effort as well as image quality. In Joseph's method and the bilinear method, the number of pixels that contribute to the estimate of the measurement is relatively constant.

A third fundamental difference is in the potential accuracy of the interpolation for the views where $|\tan \lambda|$ is close to 1. The concept of summing in x or y instead of along the ray appears sub-optimal in these views. Since they sum in x or y, Joseph's method and the distance-driven approach could be less accurate than the bilinear method, which sums along the ray. Note, however, that by allowing the summation direction to change from ray to ray within a view, Joseph's method may be less exposed to this issue than the distance-driven approach.

6.3   Experimental Setup

6.3.1   Reconstruction Method

To evaluate the effects of the projection matrix on IQ the preferences regarding the objective function and the algorithm were towards: i) a linear reconstruction strategy to facilitate resolution measurements; ii) a regularized reconstruction to avoid an excess of high frequency components in the reconstructed image; and, iii) a smooth regularization mechanism defined through the iterations to avoid interferences between the forward projection model and a Bayesian penalty term.

These requirements were best met by the Landweber algorithm as described in Section 4.3.1 without consideration of an additional statistical weighting matrix. The iteration number served as a regularization means since reconstruction was always stopped after a fixed number of iterations. The convergence controlling parameter η was chosen as 0.90 times $2/\sigma_{\max}$, where the maximum singular value of $\mathbf{A}$ was determined using five iterations of the power method [Golu 96]. In total, 1000 iterations were performed and every fifth iterate was stored, starting with the result from the first iteration. The reconstructions below 40 iterations were disregarded in the following evaluations because resolution was too low for these images to be of interest. The initial image vector, $f^{(0)}$, was always chosen as the zero vector.

6.3.2   Geometrical Settings

The comparative study included 16 different geometrical (parametric) settings that are representative of contemporary CT scans. The selection considers a moderate pixel size (MPS) and small pixel size (SPS), low and high magnification, full-scan and short-scan, and FFS off and FFS on. This variety of aspects allows to probe various features of the forward projection models. For example, as discussed earlier (Sec. 6.2.4), pixel size is an important factor for both Joseph's and the bilinear method, and high magnification may impact the distance-driven method more than the other two methods. Also, the data redundancies that are present within a full-scan may yield artifact/noise cancellations that are not feasible with a short-scan, and FFS off or on further affects data sampling.

The comparison is divided into two subsets according to the pixel size. Figure 6.6 shows the label concept that was used for the moderate and the small pixel size. Thus, each geometrical setting has an associated number together with an abbreviation denoting the pixel size. For instance, G1-MPS denotes geometrical setting 1 with moderate pixel size, where geometrical setting G1 is defined using low magnification, full-scan mode and FFS off.

All data simulation and image reconstruction parameters are summarized in Table 6.1. The grid of image pixels was always centered on the origin and the computations were only carried out over the pixels that were within the field-of-view radius, r_{FOV}, of 13 cm. The motivation of the selected pixel sizes Δ_{xy} was as follows: Modern CT scanners typically offer 512×512 images with pixels distributed over a user-selected region, the maximum length of which is 50 cm. At maximum length, the pixel size of about 0.1 cm. Thus, a pixel size of 0.075 cm would correspond to selecting a region of 38 cm, and a pixel size of 0.0375 cm to selecting a region of 19 cm. The size of 0.075 cm is representative of abdominal and chest scans of an average-sized patient, whereas the smaller size of 0.0375 cm is representative of head- or heart-focused scans. Parameters Δu and R_{FD} were chosen so that the same resolution is achieved at field-of-view center in all geometries, which is given by $(R_F/R_{FD})\Delta u$. The low magnification represents a classical head/body CT scanner, whereas the high magnification represents a head-dedicated CT scanner. The full scan covered $360°$, i.e., $[\lambda_s, \lambda_e] = [0, 2\pi)$. For the short scan, the start angle was $\lambda_s = 0$ and the end angle was $\lambda_e = 4\pi/3$. Thus, the number of projections on a short-scan was $2/3$ of that on a full-scan. When using the flying focal spot, the number of projections was doubled, as done in the scanner.

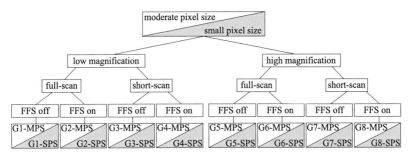

Figure 6.6: Label concept for the 16 geometric (parametric) settings.

parameter description	parameter	value	parametric case
image size	$N_x \times N_y$	351×351	MPS
	$N_x \times N_y$	701×701	SPS
image pixel size	Δ_{xy}	0.0750 cm	MPS
	Δ_{xy}	0.0375 cm	SPS
focus-detector distance	R_{FD}	104 cm	all
focus-origin distance	R_F	57 cm	low magnification
	R_F	36 cm	high magnification
detector pixel size	Δu	0.13684 cm	low magnification
	Δu	0.21667 cm	high magnification
number of detector pixels	N_u	380	all
detector pixel offset	u_{off}	1/4	FFS off
	u_{off}	1/8	FFS on
number of projections	N_λ	1200	full scan, FFS off
	N_λ	2400	full scan, FFS on
	N_λ	800	short scan, FFS off
	N_λ	1600	short scan, FFS on
number of photons	N_{ph}	$(1200/N_\lambda) \times 60,000$	all

Table 6.1: Data simulation and image reconstruction parameters. MPS, SPS and FFS are the abbreviations for moderate pixel size, small pixel size, and flying focal spot. Poisson noise was added to the data, assuming that N_{in} photons are emitted towards each detector pixel; N_{ph} is adjusted with the number of projections, so that the total exposure was the same in all geometries.

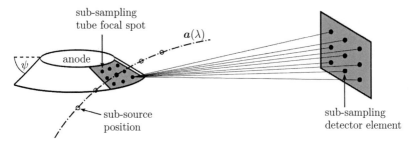

Figure 6.7: In order to simulate realistically the physical data acquisition process, a 3×3 sub-sampling of the X-ray tube focal spot, a 3×3 sub-sampling of the detector element, and five sub-source positions were used. The anode angle, ψ, was assumed to be $7°$.

6.3.3   Phantom Selection and Data Simulation

All evaluations of this study are carried out using the FORIBLD head phantom as shown in Figure 3.2. The FORBILD head phantom has been accepted as a standard in the CT community due to its simple yet challenging definition. This phantom is a three-dimensional phantom. In this work, only the central axial slice was used. The definition of that slice can be found in Yu et al. [Yu 12].

Noise-free input data for each geometrical setting was generated with a proprietary software. Data simulation was carried out in 3D, as a CT scan with one detector row of finite non-zero thickness. The program calculates the projection data in the form of analytical line integrals based on a mathematical phantom, f_{phan}, and on parameters describing the scanning geometry. In order to simulate the physical data acquisition process realistically, the simulation model included N_{fs} sub-samples of the x-ray tube focal spot, N_{de} sub-samples of the detector element, and N_{sp} sub-samples of each source position, as illustrated in Figure 6.7. These settings allowed us to model the shift-variant effect of the x-ray tube anode angle on resolution, the blurring that results from continuous x-ray emission, and the blurring that results from the finite size of the focal spot and detector elements. All sub-sampling positions were uniformly distributed. In total, N_{L} line integrals were calculated, where $N_{\text{L}} = N_{\text{fs}} \times N_{\text{de}} \times N_{\text{sf}}$. Then, the projection data of the r-th line line integral of the full simulated data set, g^S, was calculated as follows:

$$g_r^S = -\log\left(\frac{1}{N_{\text{L}}}\sum_i^{N_{\text{L}}} \exp\left(-\int_{\mathcal{L}_{r,i}} f_{\text{phan}}(\boldsymbol{x})\mathrm{d}\boldsymbol{x}\right)\right) . \tag{6.41}$$

For the present data simulation, a 3×3 sub-sampling was used for the focal spot as well as for each detector element, and each view was simulated using five sub-source positions, giving in total 405 line integrals. The x-ray tube focal spot size was $0.12\,\text{cm} \times 0.09\,\text{cm}$, and the anode angle was $7°$. For scans with FFS on, the shift was $|\delta(\lambda)| = 0.0415\,\text{cm}$ for both magnifications, as suggested in Flohr et al. [Floh 05].

All evaluations with noise involved 50 noise realizations for each geometrical setting. Poisson noise was used with a fixed number of incoming photons, N_{ph}, for each
ray. This number was changed with the number of projections to ensure that the
total exposure was always the same. Note that the number of photons was the same
for both low and high magnification, and noise simulation did not include a com-
pensating bowtie filter nor tube current modulation. All data simulation and image
reconstruction parameters are listed in Table 6.1.

6.4   Preliminary Study

This section provides the results of the preliminary study that was set up to acquire
a better understanding of how various discrete image representation techniques impact image quality. In total, seven discrete image representation techniques and four
metrics (bias, RMSE, overshoot, pixel standard deviation) were included in the investigation. In addition, some examples of the reconstructions are provided so that
the results can be partly linked with a visual impression. A summary discussion
concludes the section.

6.4.1   Study design

The preliminary study considers all forward projection models presented: the distance-
driven method, Joseph's method, the blobs, and the B-splines of order $n = 0, 1, 2, 3$.
The experimental setup as described in the previous section was applied with the
limitation that only geometrical setting G1-MPS and one noise realization was considered.

The focus was on both obtaining a visual impression of the resulting image quality
and on evaluating selected quantitative evaluation parameters. The quantitative
comparison involved an assessment of resolution, bias, RMSE, overshoot, and noise.
The evaluation is based on the basic metrics as defined in Section 5.2.

6.4.2   Visual Appearance of Some Reconstructions

Figure 6.8 shows the reconstructed noiseless and noisy images along with a respective
horizontal profile through the reconstructions after 251 Landweber iterations for all
seven forward projection models. The horizontal profile passes through the eyes and
the slanted small bones located above the rectangular bone insert in the FORBILD
head phantom. Figure 6.9 displays the position of the horizontal profile and additionally shows the noiseless reconstructions for all forward projection models after 851
Landweber iterations. The gray scale window is more compressed for the noiseless
reconstructions to emphasize the discretization errors. In addition, only the upper
part (region $y > 0$) of the reconstructed images was selected, as the other part would
not convey more information. As the images are embedded as a vector graphic, the
differences in the images are best visible when using the zoom functionality in the
PDF version.

The figures already highlight significant differences between the image representation techniques. First, each forward projection model converges at a different pace,

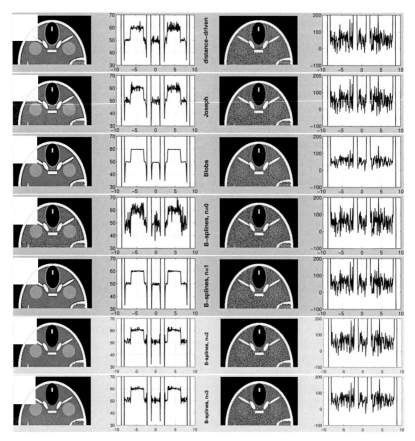

Figure 6.8: Noiseless and noisy reconstructions together with the corresponding horizontal profiles through the eyes obtained in geometry G1-MPS after 251 iterations: noise-free images *(first column)* with the respective horizontal profile *(second column)*, noisy images *(third column)* with the respective horizontal profile *(fourth column)*. From *top to bottom* the reconstructions were obtained using the distance-driven method, Joseph's method, the blobs, and the B-splines of order $n = 0, 1, 2, 3$. Noiseless images: c/w=50/40 HU; noisy images: c/w=50/200 HU. The units of the x- and y-axis of the profile plots are in units of cm and HU, respectively. The exact location of the horizontal profile is shown in Figure 6.9.

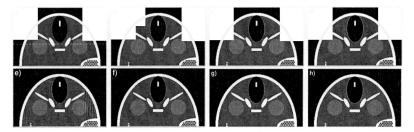

Figure 6.9: *a)* The dashed line shows the position of the horizontal profiles as displayed in Figure 6.8 in the second and fourth column. *b)-h)* Noiseless reconstructions of geometry G1-MPS after 851 iterations for the distance-driven method, Joseph's method, the blobs, and the B-splines of order $n = 0, 1, 2, 3$. Noiseless images: c/w=50/40 HU.

which is clearly visible in terms of resolution. It seems that the B-splines of order $n = 0$ converge fastest, followed by the distance-driven method, Joseph's method and the B-splines of order $n = 1$, and finally by the blobs and the B-splines of order $n = 2, 3$. Second, the noiseless reconstructions also show that the forward projection models derived from the Joseph's method and the distance-driven method yield discretization errors around the ±45-degree directions. However, the magnitude of all discretization errors is relatively small. Such discretization errors are not present at all for the forward projection models derived from the basis function approaches due to their continuous definition. Third, the over- and undershoot error varies slightly from one to the next forward projection model. This effect is expressed best in the horizontal profile plots. This impression may also be affected by the different resolutions at the given iteration steps of 251 and 851. Irrespective of the selected representations, with an increasing number of iterations, the magnitude of the over- and undershoot errors at sharp boundaries increase while reducing their spread. Also note that, in the presence of noise, most of the mentioned differences are difficult to spot.

The reconstructions with noise show that the noise pattern is slightly different between the forward projection models. For example, the noise in the reconstructions using the blob or the B-spline of order $n = 3$ looks somehow muddy, while in the other reconstructions noise grains are finer and more defined. This impression is still visible when the resolution of the reconstruction is matched, although not as strongly as before.

6.4.3 Quantitative Evaluation: Basic Metrics

Spatial resolution. As described above, slight variations in spatial resolution were observed. In order to overcome the differences in resolution, all figures-of-merit are displayed as a function of $\nu_{0.5}$; the frequency where the MTF reached a value of 0.5. Figure 6.10 illustrates the observed differences in spatial resolution over all forward projection models. The plot on the left shows the MTF curves after 201

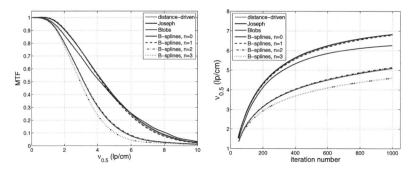

Figure 6.10: Illustration of the differences for the MTF curve for all forward projection models after 201 Landweber steps *(left)* and for the frequency at which the MTF reaches a value of 0.5 *(right)*, i.e., $\nu_{0.5}$, as a function of the iteration number.

Landweber iterations. The plot on the right shows the frequency $\nu_{0.5}$ as a function of the Landweber steps. Linear interpolation was used to bring all results on the same set of samples of $\nu_{0.5}$.

Both plots confirm the observations from the previous section. The distance-driven method, Joseph's method, and the B-splines of order $n = 1$ result in fairly similar MTF curves. The same applies to the blobs and the B-spline of order $n = 2$. The MTF curves of the B-splines of order $n = 0$ and $n = 3$ have another curve progression. The right plot clearly shows that much more iterations need to be performed for the blobs and the B-splines of order $n = 2, 3$ to reach the same resolution as, for instance, for the distance-driven method. Note that the maximum frequency that is reached after 1000 Landweber iterations is at least 4.60/cm (B-splines, $n = 3$) and in the maximum 6.83/cm (Joseph). Thus, for the blobs and the B-splines of order $n = 2, 3$ the curve in the figure-of-merit will stop earlier.

Reconstruction error. Figure 6.11 shows both the bias, σ_b, and the root mean squared error, σ_{RMSE}, as a function of $\nu_{0.5}$.

The highest bias was observed for the B-splines of order $n = 0$ with values between 4 HU and 7 HU. The bias of any other forward projection model was below 3 HU, with a slight advantage for all basis function approaches. The lowest bias for frequencies $\nu_{0.5} > 3/\mathrm{cm}$ was obtained with the blobs. Except for the B-splines of order $n = 0$, the absolute deviation of the bias between the forward projection models was relatively small - about 1 HU. The curves representing the RMSE follow the same trend as the curves of the bias.

Overshoot error. The overshoot error, σ_{os}, is shown on the left in Figure 6.12.

The overshoot error curves show altogether three different trends. First, the curve for the B-splines of order $n = 0$ runs relatively flat at about 20 HU below frequencies of 5.5/cm and increase up to 30 HU for $\nu_{0.5} > 5.5/\mathrm{cm}$. Second, for the distance-driven method, Joseph's method, and the B-splines of order $n = 1$, the overshoot error was

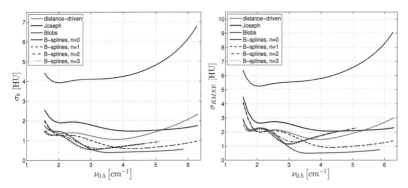

Figure 6.11: Reconstruction error measure in terms of both bias *(left)* and RMSE *(right)*, respectively, as a function of $\nu_{0.5}$.

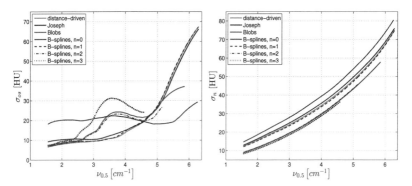

Figure 6.12: Overshoot error *(left)* and standard deviation *(right)* as a function of $\nu_{0.5}$.

below 15 HU for low frequencies ($\nu_{0.5} < 4$/cm), and then quickly increased for high frequencies of up to 65 HU. Third, the overshoot error curve for the blobs and the B-splines of order $n = 2, 3$ increased monotonously until it reached a local maximum at $\nu_{0.5} \approx 3.5$/cm, then decreased until it passed a local minimum at $\nu_{0.5} \approx 4.8$/cm, and finally increased again.

Noise.   The standard deviation, σ_n, is shown on the right in Figure 6.12.

The dependence of noise was very similar for all methods. Again, the curves may be separated into three groups: i) the B-splines of order $n = 0$; ii) the distance-driven method, Joseph's method and the B-splines of order $n = 1$; and, iii) the blobs and the B-splines of order $n = 2, 3$. The noise curve of the first group yields overall the highest noise in the image. The noise curves of the second group are vertically shifted by approximately 5 HU to lower values, and the noise curves of the third group are shifted again by approximately 5 HU in the same direction.

6.4.4 Summary Discussion

The qualitative and quantitative evaluation using basic metrics clearly showed that there are important differences in image characteristics between the forward projection models. The results can be summarized as follows:

- It seems that the characteristics of the seven forward projection models can be roughly divided into three groups: i) the B-splines of order $n = 0$; ii) the distance-driven method, Joseph's method, and the B-splines of order $n = 1$ (bilinear method); and, iii) the blobs and the B-splines of order $n = 2, 3$. Each of the group showed approximately the same patterns in terms of resolution, bias, RMSE, overshoot error and noise. Within the groups, the differences between the curve shapes were comparatively small. However, the differences across the groups were more significant.

- Although the evaluation of the computation cost was not part of the preliminary study, it may be mentioned that the reconstruction time of the Joseph's method and the distance-driven method was at least three times as fast as the basis function approaches. Furthermore, for the B-splines the reconstruction time quickly increased with increasing n.

- The forward projection models of the second group can be classified as linear interpolation models. Even though their mathematical description is fundamentally different (as described in Section 6.2.4), they seem to affect the image quality in the same way. In comparison to the third group, the linear interpolation models are in many regards more attractive. First, the computational effort needed for the reconstruction is still acceptable, while the computational cost that comes with the third group is by factors higher. Second, the discretization errors obtained with the linear interpolation models may be handled by a regularization term and in the presence of noise, these artifacts are difficult to spot. Although the third group does not have those artifacts, the convergence is quite slow. This means that this benefit is linked with a significant increase of iterations.

Finally, the preliminary study does not allow a conclusion about the best forward projection model in terms of image quality. After considering the pros and cons mentioned above, we decided to further investigate in an extensive study the effect of linear interpolation models on image quality. The results of that study are discussed below.

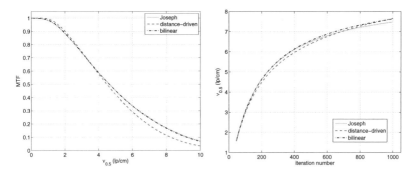

Figure 6.13: Illustration of slight variations for Joseph's method, the distance-driven method, and the bilinear method for the MTF curve after 201 Landweber steps *(left)* and for the frequency at which the MTF reaches a value of 0.5, $\nu_{0.5}$, as a function of the iteration number *(right)*. In both plots the evaluation was based on the geometrical setting G1-SPS.

6.5   Quantitative Evaluation of Linear Interpolation Models: Basic Metrics

This section provides the results obtained in a quantitative evaluation using basic metrics as described in Section 5.2. In total, eight metrics (computational cost, overshoot, bias, RMSE, mean pixel standard deviation, and mean pixel correlation coefficient in x, in y, and at 45 degrees) were included. These results are presented hereafter in a condensed format, using the abbreviations "J", "D", and "B" to refer to Joseph's method, the distance-driven method, and the bilinear method. Beforehand, some examples of reconstructions are provided so that the results can be partly linked with a visual impression. A summary discussion of the quantitative results is given after all results have been presented.

6.5.1   Presentation of the Results

6.5.1.1   Figure-of-Merit

Slight differences in spatial resolution were observed for each geometrical setting and each forward projection model. In order to overcome these effects, all figures-of-merit are displayed as a function of $\nu_{0.5}$, the frequency where the MTF reached a value of 0.5. Figure 6.13 illustrates the observed differences for the three forward projection models in the geometrical setting G1-SPS. The plot on the left shows the MTF curves after 201 Landweber iterations, and the plot on the right shows the frequency $\nu_{0.5}$ as a function of the Landweber steps. Linear interpolation was used to bring all results on the same set of samples of $\nu_{0.5}$. Note that the differences were more pronounced in the SPS geometries than in the MPS ones.

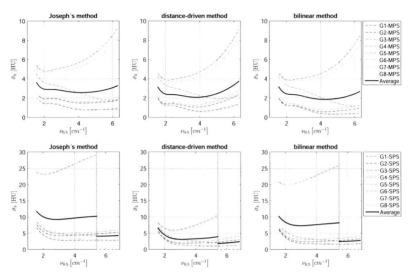

Figure 6.14: Mean absolute reconstruction error as a function of $\nu_{0.5}$ for all geometries (G1-G8) and for all forward projection models for the moderate pixel size *(first row)* and for the small pixel size *(second row)*. To minimize the number of dependencies, the averaged metric over 8 or 6 geometries is used depended on the pixel size and on the value of $\nu_{0.5}$.

6.5.1.2 Display Concept

By design, the analysis generated a large number of curves. Specifically, $3 \times 8 \times 2$ curves were obtained for each metric: 3 forward projection models, 8 geometries, and 2 pixel sizes. Results corresponding to different metrics or different pixel sizes are shown in separate figures, with the abbreviation MPS or SPS used as needed to emphasize which pixel size a figure or plot corresponds to. To minimize the number of figures, the results obtained for all three forward projection models are always incorporated together in each figure. To do so, the dependency over the 8 geometries is reduced to two summary plots.

The first plot is the metric as a function of $\nu_{0.5}$ as obtained after averaging over 6 or 8 geometries, depending on the pixel size and the value of $\nu_{0.5}$. For the moderate pixel size, the frequency range is $\nu_{0.5} \in [1.55, 6.29]$/cm and the average is always over all 8 geometries. For the small pixel size, two frequency ranges, $\nu_{0.5} \in [1.58, 5.43]$/cm and $\nu_{0.5} \in (5.43, 6.79]$/cm, are involved because the maximum frequency reached in G3-SPS and G7-SPS was lower than that reached in the other SPS cases. For the first frequency range, the average is over all 8 geometries, whereas for the second frequency range, the average is only over 6 geometries, specifically those that remain after discarding G3-SPS and G7-SPS. This approach has been thoroughly validated to avoid misleading interpretations. Figure 6.14 illustrates for all three methods and for all 8 geometries the results for the mean absolute reconstruction error as a function of $\nu_{0.5}$. The average over all geometries is highlighted as a black thick line that is discontinuous for the SPS case at $\nu_{0.5} = 5.43/\text{cm}$. Note that the average plot is sometimes driven by the geometries G3 and G7 due to the poor performance of the reconstructions.

The second summary plot is a bar plot that shows the average value of the metric over the interval $\nu_{0.5} \in [1.58, 5.43]/\text{cm}$ for each of the eight geometries. That value was placed inside angled brackets; for instance, $\langle \sigma_b \rangle_\nu$ refers to the frequency averaged mean absolute error. Note that this second summary plot is only shown for the bias, the RMSE and the noise metrics. For the other metrics, no significant differences were observed between the geometries, and therefore, this plot does not convey more information.

Three different tones of gray are represented in the following result plots: light gray for method J, medium gray for method D, and dark gray for method B.

6.5.2   Visual Appearance of Some Reconstructions

In addition to the images shown in Section 6.4, Figure 6.15 and Figure 6.16 display reconstructed noiseless and noisy images after 251 Landweber iterations for method J, D, and B in geometry G1-MPS and G1-SPS, respectively. Note that the gray scale window is more compressed for the noiseless reconstructions to emphasize the discretization errors. Also, only the upper part (region $y > 0$) of the reconstructed images is shown, as the other part would not convey more information.

As expected, method J and D yield discretization errors around the ± 45-degree direction which can be clearly recognizable in the noiseless reconstructions. However, the magnitude of all discretization errors is relatively small. In the presence of noise, the discretization errors are difficult to see, and so are other possible differences between the three forward projection models.

In G1-SPS, methods J and B both yield more discretization errors than method D, but the magnitude of the discretization errors remain relatively small, so that it is still difficult to distinguish these differences in the presence of noise. Unlike in G1-MPS though, method D visibly yields a different noise structure in comparison with the other two methods.

6.5.3   Results

Computational effort.   In our non-optimized implementation, the computation times of the matrix elements a_{rs} varied from method to method, whereas method B required significantly more effort, primarily because no look-up table was used. Also, computing the elements of $\mathbf{A}$ took more time with method D than with method J, particularly for the SPS geometries.

Table 6.2 lists the adjustment factors (Eq. 5.2) for each method. Since method J always had the lowest number of non-zero elements in $\mathbf{A}$, the adjustment factor becomes 1.00. Note that the adjustment factor of method B is about the same for both pixel sizes, whereas the adjustment factor of method D increase from 1.05 to 1.60 for SPS, since more image pixels overlap with the interpolation interval $[q_1, q_2]$.

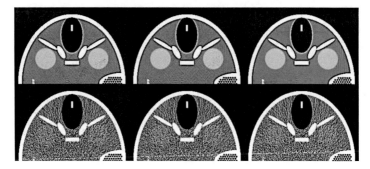

Figure 6.15: Noiseless and noisy reconstructions obtained in geometry G1-MPS after 251 Landweber iterations: method J (*first column*), method D (*second column*), and method B (*third column*). Noiseless images: c/w=50/40 HU; noisy images: c/w=50/200 HU.

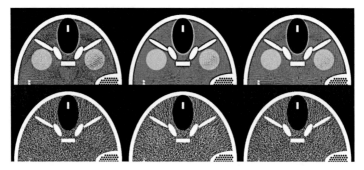

Figure 6.16: Same reconstructions and arrangement as in Figure 6.15 but for geometry G1-SPS.

Figure 6.17 compares the three methods in terms of computational cost as described in Section 5.2.2. Method J always required less effort. The increase in effort for method B varied between 45% and 70%, and, for method D, the increase was strongly dependent on the pixel size: it was only around 4% for MPS, but varied between 51% and 73% for SPS.

Overshoot error.   The overshoot error, σ_{os}, for all methods is shown in Figure 6.18. For MPS, the overshoot error followed the same trend for all methods. For low frequencies ($\nu_{0.5} < 4$/cm) the overshoot error was below 15 HU, but σ_{os} increased quickly for high frequencies. At maximum frequency, the overshoot error was slightly above 65 HU for all methods, with a difference of about 10% in favor of method D.

For SPS, the curve $\sigma_{os}(\nu_{0.5})$ of method J and B was significantly different compared to method D. The overshoot error of method D followed the same trend as in MPS,

adjustment factor	method J	method D	method B
σ_{A-MPS}	1.00	1.05	1.52
σ_{A-SPS}	1.00	1.60	1.53

Table 6.2: Adjustment factor for each forward projection model to account for differences in the number of non-zero elements in **A**. For both pixel sizes, method J yields the lowest number of non-zero elements.

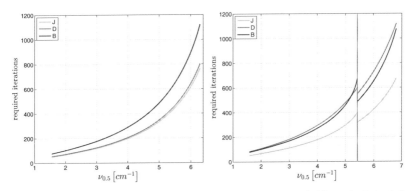

Figure 6.17: Computational cost as a function of $\nu_{0.5}$ for MPS *(left)* and SPS *(right)*. Only the first summary plot is shown.

whereas the overshoot error for methods J and B essentially stayed the same over all frequencies. The overshoot error was always below 20 HU for methods J and B, and always about 5 HU lower for method B compared to method J. At low frequencies ($\nu_{0.5} < 4$/cm), the error for method D was slightly smaller than that for method B by about 2.5 HU. At maximum frequency, the error for method D was about 37 HU, which is comparatively high, but 43% less than for MPS.

Reconstruction error.   Figures 6.19 and 6.20 compare the methods in terms of bias and root mean squared error.

For MPS, the geometry-averaged bias was always below 4 HU, with a slight advantage in favor of method B. Also, the bias in geometries 3 and 7, which are the short-scans without FFS, was about twice larger for all methods. The curves for the root mean squared error followed the same trend as the curves of the bias, but were shifted off about 1.5 HU to higher values.

For SPS, the bias was below 4-5 HU for all three methods, except in geometries 3 and 7, where it rose above 20 HU for methods B and J, while staying below 8 HU for method D. Poor performance in geometries 3 and 7 drove the major differences observed in the geometry-averaged bias for $\nu_{0.5} < 5.43$/cm. Above this frequency, these two geometries were discarded and the bias fell below 5 HU for all methods, with methods B and D performing better than method J by about 1.5 HU and 2 HU,

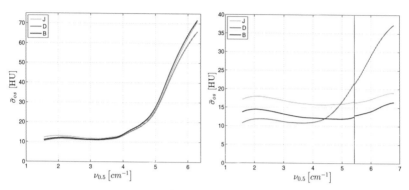

Figure 6.18: Overshoot as a function of $\nu_{0.5}$ for MPS *(left)* and SPS *(right)*. Only the first summary plot is shown.

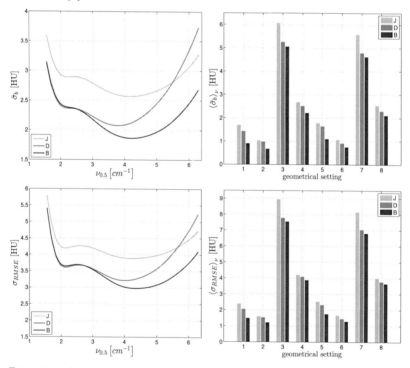

Figure 6.19: Reconstruction error measures for MPS in terms of both bias *(top row)* and RMSE *(bottom row)*. Geometry-averaged bias and RMSE, respectively, as a function of $\nu_{0.5}$ *(left)*. Frequency-averaged bias and RMSE, respectively, for each geometry *(right)*.

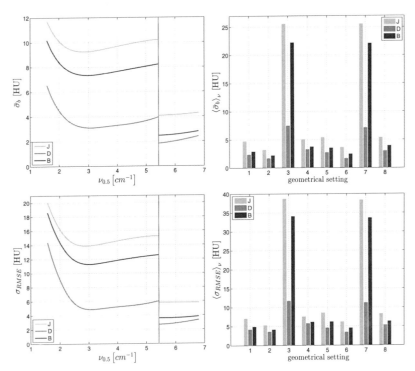

Figure 6.20: Reconstruction error measures for SPS in terms of both bias *(top row)* and RMSE *(bottom row)*. Geometry-averaged bias and RMSE, respectively, as a function of $\nu_{0.5}$ *(left)*. Frequency-averaged bias and RMSE, respectively, for each geometry *(right)*.

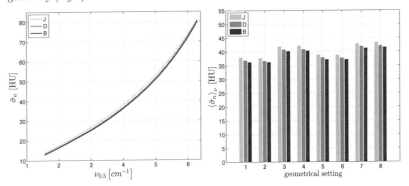

Figure 6.21: MPS results for the mean standard deviation for the geometry-averaged curves *(left)* and the frequency-averaged values *(right)*.

respectively. As in MPS, the root mean squared error follows again the same trend as the bias, but is again shifted to higher values of σ_{RMSE}.

Noise. Figures 6.21 to 6.25 compare the methods in terms of noise performance for MPS and SPS, respectively.[1]

For MPS, the dependence of noise metrics on frequency and geometry was very similar for all methods. Differences of about 2 HU were observed in the mean standard deviation, with methods B and J always performing best and worst, respectively. In addition, there were significant differences in the correlation coefficient: method J yields less correlation between the pixel values than methods D and B, with differences as high as 0.05 and 0.1 compared to methods D and B, respectively.

For SPS, the dependence of noise metrics on frequency and geometry was also observed to be very similar for all methods. However, differences as large as 40 HU were observed in the mean standard deviation, and methods D and J became the methods that always perform best and worst, respectively. Moreover, the differences in the correlation coefficient dramatically increased. A difference as large as 0.3 was observed between methods J and D. Also, unlike in MPS, method B created correlation levels that are much closer to those of method J than method D.

6.5.4   Summary Discussion

The quantitative evaluation using basic metrics clearly showed that there are important differences in image characteristics for method J, D, and B. The differences turned out to be more significant in the SPS case in particular. The results can be summarized as follows:

- The overshoot error turned out to yield interesting differences for the small pixel size whereas the moderate pixel size showed nothing conspicuous. In SPS, the overshoot for method J and B was found to be relatively stable over the frequency range, while the overshoot error increased with resolution for method D. Which overshoot behavior is preferable is, however, not clear and most likely task-dependent, since overshoot is sometimes intentionally induced in clinical protocols to enhance edges.

- Method J and D are weak in their description of lines around the ±45-degree directions, which become visible in artifacts along that direction, whereas method B avoids such artifacts. Thus, method B yield the lowest reconstruction error (bias/RMSE) for MPS with the drawback of large computation effort ($> 40\%$) compared to the other methods. In the SPS case, method B could be seen as an improvement over method J in terms of bias, but, again, the computational cost for that was high ($> 40\%$). For method D, the artifacts around the ±45-degree directions seemed to be blurred due to the increasing integration lengths, which let the reconstruction error drop below method B. However, that gain was accompanied with a high computational cost ($> 50\%$).

---

[1]Note that the statistical accuracy of the mean standard deviation was very high. Specifically, a utilization of another set of 50 noise realizations in geometry G1 yields less than 1% difference in the mean standard deviation. This accuracy was assumed to be similar for all geometries.

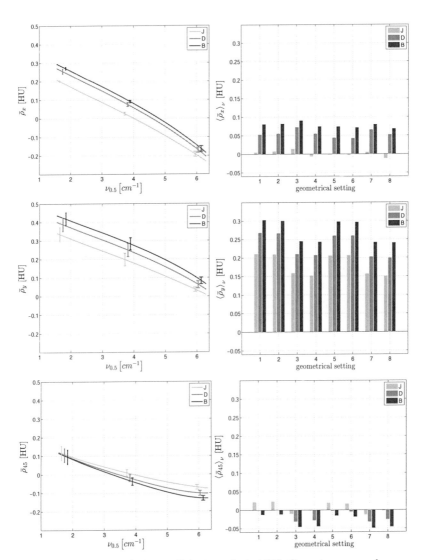

Figure 6.22: Mean correlation coefficient results in MPS. Geometry-averaged curves *(left)*. Frequency-averaged values *(right)*. From top to bottom: correlation coefficient in x, y, and in 45 degree.

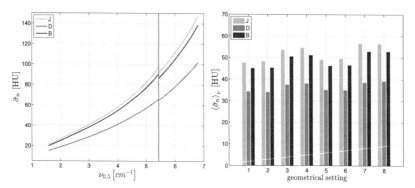

Figure 6.23: SPS results for the mean standard deviation for the geometry-averaged curves *(left)* and the frequency-averaged values *(right)*.

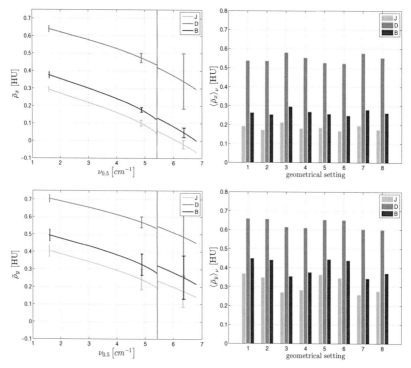

Figure 6.24: Mean correlation coefficient results in SPS. Geometry-averaged curves *(left)*. Frequency-averaged values *(right)*. From top to bottom: correlation coefficient in x, and y.

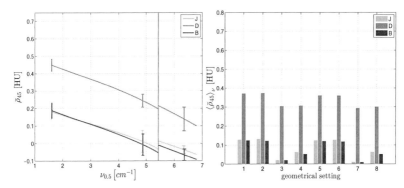

Figure 6.25: Mean correlation coefficient results in SPS. Geometry-averaged curves *(left)*. Frequency-averaged values *(right)*. From top to bottom: correlation coefficient in 45 degree.

- In terms of reconstruction error induced by using small pixels, method D is more robust than methods B and J. However, the bias was remarkably low in many settings, so that the bias advantage of method D may only be relevant in particular cases. In our study, these cases were G3-SPS and G7-SPS, which are the only two cases for which the reconstruction problem was under-determined. Unfortunately, this advantage of method D comes with a major increase in computational effort ($> 50\%$).

- In terms of noise, it is difficult to single out a preferred method since the the mean standard deviation behaves contrary to the the mean correlation coefficients. For instance, the pixel standard deviation was always larger for method J, but this effect was systematically accompanied by lower correlations between the pixels. The pixels in method D were much more correlated, which is not surprising. By including the size of the detector pixel, method D simply enforced more binding between the pixels.

The summary shows that the observed differences were not such that one method can straightforwardly be identified as being in favor over all others in all settings. Task-based assessment of image quality is needed to identify how the observed trade-off effects play out.

6.6   Quantitative Evaluation of Linear Interpolation Models: Task-based Metrics

As observed in the previous section, the linear forward projection models yield differences and no method could be claimed to be preferable. Task-based assessment of image quality is needed to further rank the methods. The task-based assessment presented here involved an ideal observer study, a CHO study, and a human observer study.

6.6.1   Ideal Observer Study

6.6.1.1   Task Selection

To evaluate the differences in image quality, the ideal observer was applied to a SKE/BKE binary classification task. The AUC was used to measure the performance of the ideal observer.

Let Class 1 contain all images where the signal is normal, and let Class 2 contain all images where the signal corresponds to a disease. Class 1 and Class 2 images were sought to be representative of a medical task. Thus, the Class 1 signal was defined as a circle of radius r_c and center point C. Then, the Class 2 signal was created from the Class 1 signal as follows. First, the Class 1 signal was truncated, using four clipping lines at a distance of $\pm d_c$ of C in the vertical and horizontal directions. The remaining part of the signal looks like a square with blunt edges. Four circles of radius r_l at distance $\pm d_l$ from C were superimposed on this square to yield the Class 2 signal. The Class 1 image, the procedure to obtain the Class 2 image, and an example of a possible Class 2 image is shown in Figure 6.26. This definition of the Class 2 image yields various expressions by varying d_l and r_l, while using a fixed value for r_c and d_c. The involved values were $d_l = 0.30, 0.34, \ldots, 0.50$cm, $r_l = 0.25, 0.27, \ldots, 0.37$cm, $r_c = 0.65$cm and $d_c = 0.55$cm. Figure 6.27 shows the obtained Class 2 images for each combination of d_l and r_l. The medical analogy to that selection of classes is the classification between lesions with a fuzzy or sharp boundary, as fuzziness is often a marker for malignancy.

To find a preferred setting for the Class 2 signal, 10000 noisy FBP reconstructions were performed at three resolution levels in geometry G1-MPS. In each class, the difference between the signal and the background was 10 HU and the lesion center C was placed in the central low contrast ellipse of the FORBILD head phantom at location $(x_l, y_l) = (0.0, -3.6)$. Then, the AUC value $\mathcal{A}$ corresponding to the various expressions of the Class 2 signal was computed. For this computation, all pixels that were at a distance of $r_{LROI} = 8.25$ mm or less from C were involved in the definition of μ_1 and μ_2. Only small differences were observed when changing the resolution of the reconstruction. The results of this preliminary experiment are shown in Table 6.3 for a reconstruction with no apodization on the ramp filter. Note that in four cases, the total lesion size became bigger than the region that was used for the calculation of $\mathcal{A}$. Thus, for those cases, it was not possible to calculate an AUC value. Based on the results in in Table 6.3, the signal corresponding to $d_l = 0.34$ cm, $r_l = 0.27$ cm and $\mathcal{A} = 0.712$ was selected for Class 2.

Any further evaluations based on that signal selection involved p pixels that are at distance $r_{LROI} = 8.25$ mm or less from C, which yield $p = 373$ in MPS and $p = 1513$ in SPS.

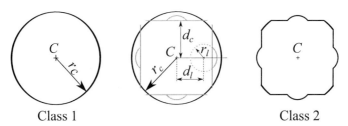

Class 1   Class 2

Figure 6.26: The Class 1 image is a circle with radius r_c *(left)*. Procedure to create a Class 2 image *(middle)*. An example of a class 2 image (lesion object) *(right)*.

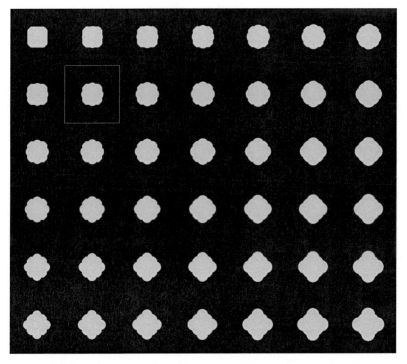

Figure 6.27: Illustration of possible Class 2 images following the procedure illustrated in Fig. 6.26. Variation of the lesion radius $r_l = 0.25, 0.27, \ldots, 0.37$ cm *(left to right)*. Variation of the distance d_l; $d_l = 0.30, 0.34, \ldots, 0.50$ cm *(top to bottom)*. The white bordered lesion was selected for ideal observer study.

		r_l [cm]						
		0.25	0.27	0.29	0.31	0.33	0.35	0.37
d_l [cm]	0.30	0.792	0.777	0.748	0.706	0.653	0.594	0.560
	0.34	0.751	0.712	0.663	0.609	0.572	0.601	0.667
	0.38	0.675	0.625	0.586	0.599	0.655	0.720	0.782
	0.42	0.604	0.602	0.645	0.704	0.764	0.818	0.863
	0.46	0.639	0.690	0.746	0.799	0.845	0.882	−−
	0.50	0.728	0.779	0.825	0.863	−−	−−	−−

Table 6.3: AUC values, $\mathcal{A}$, obtained using 10000 FBP noise reconstruction with no apodization. In four cases, marked with −−, the lesion became bigger than the circular region that was used for the AUC calculation.

6.6.1.2 Controlling Statistical Variability

To control and limit the statistical variability, Equations 5.14 and 5.15 were used to calculate the AUC value. However, the number of measurements required for Classes 1 and 2 together, $m+n$, needs to be larger than p, which is particularly challenging given the above values of p. To overcome this problem while using (only) 50 noise realizations for each geometry (which already entail significant computer resources), the following approach was adopted. Each noise realization was considered to provide 49 samples instead of 1 sample for the computation of the sample covariance matrix $\mathbf{S}$. To obtain the 49 samples, the noise properties were assumed to vary slowly around the signal location. The first sample was obtained, as expected, using the p pixels centered on C. Then, the other 48 samples were obtained using p pixels centered on shifted locations. These locations were on a 7×7 grid centered on C with a pitch, p_{LROI}, of 18 mm. Figure 6.28 shows on the left the analytical FORBILD head phantom with a uniform, circular lesion (Class 1) together with the 49 LROIs. The sampling parameters of the LROIs are illustrated on the right of this figure. With that approach, the total number of samples involved in AUC computation was 2450.

To validate the concept described above, the 10000 FBP reconstructions were used to compute the AUC once by using 1 and once by using 49 samples per noise realization. The result was an $\mathcal{A} = 0.7251$ for the approach with 49 samples, and $\mathcal{A} = 0.7122$ for the other case. This result provided enough confidence for utilization of the approximation.

At the same time, to determine how reliable formula 5.17 would remain to evaluate statistical variability, the 10000 FBP reconstructions were split into 200 sets of 50 noise realizations, yielding each one AUC value. The standard deviation over these 200 values was computed and compared to the mean value obtained from Equation 5.17. The result was a sigma value of 0.00295 in the first case versus 0.00310 in the second case, demonstrating excellent agreement.

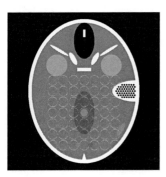 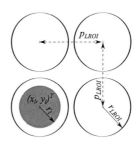

Figure 6.28: Analytical FORBILD head phantom with a uniform, circular lesion of radius r_c of 10 HU contrast (display window: [0 100] HU). The white bordered areas are the circular LROIs that were used for the calculation of the sample pooled covariance matrix $\mathbf{S}$ *(left)*. Illustration of the sampling parameters of the circular LROIs *(right)*.

Note that these tests were performed in setting G1-MPS with no apodization on the ramp filter. No significant differences are expected for different settings and resolution level.

6.6.1.3   Results

Figure 6.29 shows the noiseless image reconstructions of the Class 1 and a Class 2 image after 251 Landweber reconstructions for geometry G1-MPS using method B. The images represent an area of size 15×15 mm^2. In the reconstructions with noise, the human eye cannot distinguish the shape of the lesion any longer.

The AUC results are shown in exactly the same two formats as in Section 6.5. Figures 6.30 and 6.31 show the AUC results for all methods in the ideal observer study.

For MPS, there are practically no differences in the performance between the three forward projection models. The standard error associated with these plots was as follows: In the first summary plot, the standard deviation value for each point on the curves was about 0.0011 for each forward projection model. In the second format, the standard deviation on each bar was about 0.0031. These numbers convey high statistical accuracy and further convey that the small differences observed in each format are not statistically significant.

For SPS, there is again practically no difference in performance between the three forward projection models. The standard error associated with these plots was as follows: In the first format, the standard deviation value for each point on the curves was about 0.0016 for each forward projection model. In the second format, the standard deviation on each bar was about 0.0046. As in the MPS case, these numbers convey high statistical accuracy, and also that the small differences observed in each format are not statistically significant.

Figure 6.29: Noiseless reconstructions obtained in geometry G1-MPS after 251 Landweber iterations using method B. Cropped Class 1 image *(left)*. Cropped Class 2 image *(right)*. Only a small area around the lesion is shown (15×15 mm^2). Display window: c/w=50/100.

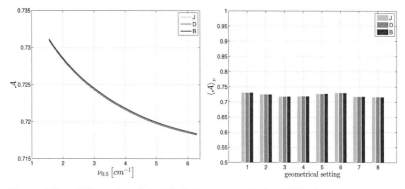

Figure 6.30: AUC results of the ideal observer study for MPS. Geometry-averaged AUC curves *(left)*. Frequency-averaged AUC values *(right)*.

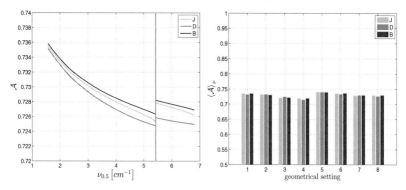

Figure 6.31: AUC results of the ideal observer study for SPS. Geometry-averaged AUC curves *(left)*. Frequency-averaged AUC values *(right)*.

The results show that the choice of the model has very little influence on the performance of a task. The differences observed in pixel noise variance and correlation seem to play no role for the ideal observer.

6.6.2   Channelized Hotelling Observer Study

6.6.2.1   Task Selection

The channelized Hotelling observers were used to evaluate IQ for a lesion detection task. In total, the same 40 Gabor channels were used for both pixel sizes with the parameter settings given in Section 5.3.3.2. The identical Class 1 and Class 2 images, including shape, size, and contrast, as in the ideal observer study (Sec. 6.6.1) were used. The CHO performance was measured in terms of AUC without and with 25% internal noise, i.e., $\xi = 0\%$ and $\xi = 25\%$.

6.6.2.2   Controlling Statistical Variability

The same approach of making use of several LROIs as described in Section 6.6.1.2 was applied to control and limit the statistical variability. For the CHO study, the shape, size, and number of the LROIs was changed. Instead of circular LROIs, squared LROIs were used, since the Gabor channels are usually defined for a square or rectangular image size. The squared LROIs were of size 23.25×23.25 mm^2 and the center to center distance from box to box was 29.25 mm. In total, 25 samples were selected where the samples were located on a 5×5 grid centered on the lesion location. Equations 5.19 and 5.15 were used to calculate the AUC point estimator. Note that using Equation 5.15 in the CHO approach means that the parameter p is equal to the number of channels, i.e., $p = 40$.

6.6.2.3   Results

Figures 6.32 and 6.33 show the AUC results for all methods without and with 25% internal noise in MPS and SPS geometry, respectively.

For MPS, there are practically no differences in the performance between the linear interpolation models without and with internal noise. The curve $\mathcal{A}(\nu_{0.5})$ without internal noise lies about 0.02 above the curve with 25% internal noise, for instance at $\nu_{0.5} = 4.64$/cm the AUC value was $\mathcal{A}_0 = 0.6994$ and $\mathcal{A}_{25} = 0.6766$. Only the second summary plot for $\xi = 0\%$ is shown since the plot for $\xi = 25\%$ does not convey any further information. The standard error associated with these plots was as follows: In the first summary plot, the standard deviation value for each point on the curves and for each forward projection model was about 0.0013 and 0.0012 for $\xi = 0\%$ and $\xi = 25\%$, respectively. In the second summary plot, the standard deviation on each bar was about 0.0037 and 0.0033 for $\xi = 0\%$ and $\xi = 25\%$, respectively. These numbers convey high statistical accuracy, and further convey that the tiny differences observed in each format are not statistically significant.

For SPS, there is again practically no difference in performance between the three forward projection models. The curve $\mathcal{A}(\nu_{0.5})$ without internal noise lies again about 0.02 above the curve with 25% internal noise, for instance at $\nu_{0.5} = 4.64$/cm the

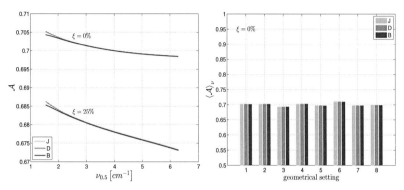

Figure 6.32: AUC results of the CHO study for MPS without and with 25% internal noise. Geometry-averaged AUC curves *(left)*. Frequency-averaged AUC values *(right)*.

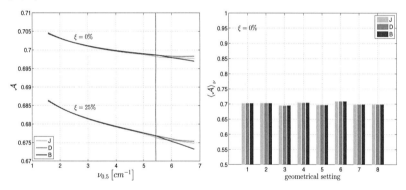

Figure 6.33: AUC results of the CHO study for SPS without and with 25% internal noise. Geometry-averaged AUC curves *(left)*. Frequency-averaged AUC values *(right)*.

AUC value was $\mathcal{A}_0 = 0.6993$ and $\mathcal{A}_{25} = 0.6783$. Only the second summary plot for $\xi = 0\%$ is shown since the plot for $\xi = 25\%$ does not convey any further information. The standard deviation value for each forward projection model for each point in the first summary plot was about 0.0013 and 0.0012 for $\xi = 0\%$ and $\xi = 25\%$, respectively. In the second summary plot, the standard deviation on each bar was about 0.0037 and 0.0033 for $\xi = 0\%$ and $\xi = 25\%$, respectively. As in the MPS case, these numbers convey high statistically accuracy, and further convey that the tiny differences observed in each format are not statistically significant.

The CHO study shows that the choice of the model has very little influence on the performance of a task. As expected, the AUC values dropped about 0.02 to 0.05

in comparison to the ideal observer study. The differences observed in pixel noise variance and correlation seem to play no role for the channelized Hotelling observer.

6.6.3   Human Observer Study

Based on the results obtained in the ideal observer study and in the CHO study, we hypothesize that the differences observed in noise (see Sec. 5.2) balance out in the context of performing tasks with the image. This hypothesis was tested using LROC analysis with human observers.

6.6.3.1   Task Selection

Four human observers were asked to perform a 2-AFC experiment. The first two observers, called Reader A and B hereafter, were CT scientists (namely, F. Noo and K. Hahn), whereas the other two observers, called Reader C and D, were neuro-radiologists (H. C. Davidson and U. Rassner). The task was to detect a circular lesion that was placed on a random location within the background of the FORBILD head phantom, with no overlap with the air and bone regions. The lesion had a fixed diameter of 0.70 cm and a random contrast between 23 HU and 33 HU. The figure-of-merit for the LROC analysis was the area under the curve, called $\mathcal{A}_L$.

Note that the results of the previous two studies served as a guide for the design of the lesion with the aim to reach an AUC value of about 0.80. To make the task for the readers a little easier, the lesion size was chosen to be circular. Alternatively, the lesion shape of the Class 2 images as defined in Section 6.6.1 and displayed on the right in Figure 6.29 would have been conceivable. In comparison to the previous sections, the contrast of the lesion was increased in order to prevent a rapid fatigue of the readers and to obtain an AUC that is around the target value.

The experiment was exclusively performed for the SPS geometries for two reasons: i) the observed differences in noise in the quantitative assessment using basic metrics (Sec. 6.5) were smaller for the MPS than for the SPS geometries; and, ii) the results obtained in previous model observer experiments (Sec. 6.6.1 and 6.6.2) were for both geometries not statistically relevant; thus, differences may be expected for SPS. Further, the geometrical settings G3 and G7 were excluded. Then, in each noise realization, the iterate that yields a resolution of $\nu_{0.5} = 4.64/\text{cm}$ was selected for the study. Thus, 300 reconstructions were available for the human observer experiment.

In order to increase the number of cases and thereby mitigate statistical variability, each reconstruction was split into two halves. In addition, to mitigate statistical dependence a small border region was removed. Thereby, the study involved 50×6 pairs of images (600 half-images) for each method for both training and testing. The images were shuffled using uniform random generators. This yield altogether four possible image pairs that were presented to the reader as shown in Figure 6.34. The display window was fixed at a window level of 50 HU and a window width of 200 HU. One and only one of the two presented images contained exactly one lesion whereas the other image was lesion-free.

Each observer read 300 pairs of images in one session. A session consisted of three sub-sessions, each of which was for one method. Each sub-session started with a training set that included 36 image pairs to gain prior knowledge about the lesion characteristics. The testing session involved 264 image pairs and was started immediately after the training session. The ordering of the sub-sessions was also randomized from one reader to the next. All observers read the images in the same dimmed dark room at an illuminance level of 10 lux. The images were displayed on a medical grade monitor that was calibrated following the ACR Technical Standard for Electronic Practice [Norw 13]. The reading distance was approximately $40-50$ cm and the reading time was approximately one hour per sub-session. After completing one sub-session each observer was asked to take a short break of about 15 minutes to avoid fatigue.

6.6.3.2   Results

Figure 6.35 shows on the left side the mask that was used to randomly distribute the lesions within the background of the FORBILD head phantom. On the right in that figure, the true lesion locations with the corresponding true, randomly determined contrast are displayed. Each half-image contain 132 lesion locations.

To compute the area under the LROC curve, a suitable localization radius is important. The localization radius is a criterion that is used to decide if a lesion is deemed to be correctly localized or not. If the observer set the mark within a certain distance to the lesion center, the lesion is found to correctly localized. To identify a good localization radius, first the correctly localized fraction of lesions was plotted as a function of the localization radius. Then, a localization radius is chosen such that the value where the rate of change in that fraction is small across all available curves. This procedure is the same as in Gifford et al. [Giff 07] and Kadrmas et al. [Kadr 09].

Figure 6.36 shows the fraction of lesions found as a function of the localization radius in pixel units for all readers and all methods. The bold black line in this plot indicate the average over all curves. Based on that plot, a lesion was deemed to be correctly localized if the mark was set within 12 pixels from the lesion center. Note that the total lesion radius in pixels units was 18.

Table 6.4 summarizes both the AUC values for each individual reader and the reader average together with the associated standard deviations. To define the reader average, let m denote the method ($m = J, D, B$), let r denote the reader ($r =$ A,B,C,D), let $\mathcal{A}_L^{m,r}$ be the individual reader performance of reader r for method m, and let $\bar{\mathcal{A}}_L^m$ be the reader average of method m. Then, the reader average for each method is given by

$$\bar{\mathcal{A}}_L^J = (\mathcal{A}_L^{J,A} + \mathcal{A}_L^{J,B} + \mathcal{A}_L^{J,C} + \mathcal{A}_L^{J,D})/4 \ ,$$
$$\bar{\mathcal{A}}_L^D = (\mathcal{A}_L^{D,A} + \mathcal{A}_L^{D,B} + \mathcal{A}_L^{D,C} + \mathcal{A}_L^{D,D})/4 \ ,$$
$$\bar{\mathcal{A}}_L^B = (\mathcal{A}_L^{B,A} + \mathcal{A}_L^{B,B} + \mathcal{A}_L^{B,C} + \mathcal{A}_L^{B,D})/4 \ .$$

The associated confidence regions were identified using the statistical procedure described in Noo et al. [Noo 13].

The image sets used in that LROC experiment were statistically dependent. To further compare the correlated results obtained from the LROC observer experiment, the vector of the figure-of-merit was chosen to be $\boldsymbol{d} = [\bar{\mathcal{A}}_L^J, \bar{\mathcal{A}}_L^J - \bar{\mathcal{A}}_L^D, \bar{\mathcal{A}}_L^J - \bar{\mathcal{A}}_L^B]$. The confidence intervals were calculated following the statistical approach described

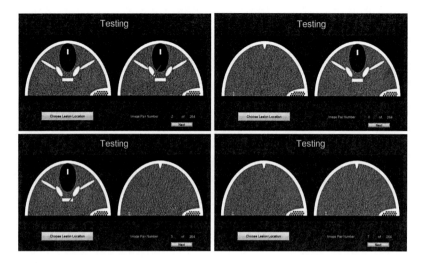

Figure 6.34: Possible image pairs presented to the reader within a testing session. In each case the lesion is present in one and only one of the two images. The observer was asked to select one lesion location within one of the two images. The lesion location in these examples is indicated by arrows. c/w = 50/200 HU.

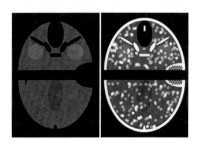

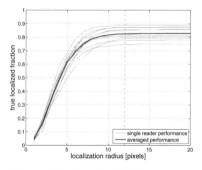

Figure 6.35: Mask within the lesions were randomly distributed *(left)*. True lesion locations as they were used in the human observer study. They are illustrated in the original noiseless phantom; each half includes 132 lesions *(right)*. c/w=50/100 HU.

Figure 6.36: True localized fraction as a function of the localization radius for all readers and all methods. The bold black line indicates the average of all curves, and the dashed line indicates the radius that within each lesion was deemed to be correctly localized.

	Reader A	Reader B	Reader C	Reader D	Reader-average
J	0.890 ± 0.019	0.883 ± 0.020	0.777 ± 0.026	0.818 ± 0.024	0.842 ± 0.017
D	0.875 ± 0.020	0.886 ± 0.020	0.792 ± 0.025	0.849 ± 0.022	0.850 ± 0.017
B	0.871 ± 0.021	0.867 ± 0.021	0.746 ± 0.027	0.826 ± 0.023	0.828 ± 0.017

Table 6.4: Areas under the LROC obtained in the human observer study with the corresponding standard deviation for each reader and for each method. Again, the abbreviations "J", "D", and "B" refer to Joseph's method, the distance-driven method, and the bilinear method.

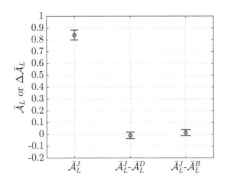

Figure 6.37: Point estimates, denoted as circles, and confidence intervals for $\bar{A}_L^J$, $\bar{A}_L^J - \bar{A}_L^D$, and $\bar{A}_L^J - \bar{A}_L^B$. The coverage probability was selected to be 98.33%, so that the joint coverage probability for the three intervals is at least 95%.

in Noo et al. [Noo 13]. More specifically, for the first entry in $\boldsymbol{d}$, an asymmetric confidence interval was estimated with the logit-transformation approach advocated by Pepe [Pepe 03, p. 107]. For the other two entries in $\boldsymbol{d}$, symmetric confidence intervals were estimated with the conventional Wald method [Pepe 03, p. 107]. A coverage probability of 98.33% was selected for each confidence interval. Then, by Bonferroni's inequality[2], the joint probability for the three intervals together is at least 95%.

Figure 6.37 shows the point estimates with their associated confidence intervals for $\boldsymbol{d}$, where the point estimates are indicated with circles. The numerical values were as follows: $\bar{A}_L^J = [0.800, 0.884]$, $\bar{A}_L^J - \bar{A}_L^D = [-0.036, 0.019]$ and $\bar{A}_L^J - \bar{A}_L^B = [-0.012, 0.040]$. Examining the results, no statistical significant difference is observed between the methods. Thus, there is no evidence to reject the null hypothesis that a difference in performance is achieved using one method against another.

---

[2]For arbitrary events $E_1, E_2, \ldots, E_n$, the Bonferroni inequality [Case 01, Eq. 1.2.10, p. 13] takes the form $P\left(\bigcap_{i=1}^{n} E_i\right) \geq \sum_{i=1}^{n} P(E_i) - (n-1)$, where $P(\Gamma)$ stands for probability of event Γ [Wund 12a, Noo 13].

6.6.4   Summary Discussion

The quantitative evaluation using task-based metrics clearly showed that small to tiny differences in the AUC measure may be observed. However, in all cases these differences are not statistically significant. The results can be summarized as follows:

- Small differences between the methods for SPS were observed in the ideal observer study. These differences were less than 0.002 and therefore statistically not significant. For MPS the observed differences were much smaller and further not statistically significant.

- By using a more human-like model observer model, the differences observed in the ideal observer study became even smaller. In the CHO study with and without internal noise, the AUC curves for both geometries are very close to each other and no method could be claimed to be preferable to the other since the differences were not statistically significant.

- The SPS geometries were under further investigation in an LROC experiment with human observers. That study also shows that the AUC performance of all methods was comparable, where method D received the highest reader average, closely followed by method J and B. Again, these differences were not statistically significant.

6.7   Discussion and Conclusion

The motivation of this study was to identify the impact of discrete forward projection models on image quality. Such an assessment is helpful to gain the most out of iterative reconstruction methods. By knowing beforehand if and, if so, how, the forward projection model may influence the IQ will allow to optimize the design of an iterative reconstruction algorithm. Therefore, a preliminary and two extended studies were set up.

In the preliminary study, in total seven forward projection models were under investigation. The comparison was based on a single geometrical setting as the primary goal was to better understand the impact of various forward projection models on image quality. In this study, the focus was on a qualitative and quantitative evaluation. The quantitative comparison involved an assessment of resolution, bias, RMSE, overshoot, and noise.

The results of the preliminary study showed that the resulting image quality from the different forward projection models can be roughly subdivided into three groups. The best image quality in terms of bias and noise was observed for the third group, which consisted of forward projection models that were only based on a basis function approach, specifically the blobs and the B-splines of order $n = 2, 3$. The third group was closely followed by the second group, which was represented by linear forward projection models. While the performance of bias and noise properties was a little worse than that of the first group, higher resolutions were reached much more quickly. For the first group, the bias and noise properties were unattractively high while the convergence was not faster than that of group two.

Thus, the following considerations were taken into account for the extended study. The B-splines of order $n = 1$ showed non-preferred image quality and were excluded in further image quality assessment. A serious drawback of the forward projection models of the blobs and the B-splines of order $n = 2, 3$ was the very slow convergence. In our non-optimized implementation, the reconstruction time of the other methods was at least three times faster. These aspects have raised the question of which of the remaining forward projection models are preferable.

Hence, three popular forward projection models were under investigation for the two extended studies: Joseph's method, the distance-driven method, and the bilinear method. An extensive investigation of these three models has been conducted including a quantitative evaluation using basic and task-based metrics. The comparison was extensive in terms of geometrical settings, including cases that challenge each model. The evaluation using basic metrics covered spatial resolution, computational cost, overshoot cost, reconstruction error, and noise measurements. The evaluation using task-based metrics was based on the AUC analysis and was carried out for an ideal observer study, a CHO study without and with internal noise, and an LROC experiment with human observers.

The evaluation using basic metrics showed that there are important differences in image characteristics for Joseph's method, the distance-driven method, and the bilinear method. In two geometrical settings, G3-SPS and G7-SPS, the reconstruction error induced by using small pixels was significantly smaller ($> 65\%$) for the distance-driven method than for the other two methods. These two short-scan cases were the only two settings for which the reconstruction problem was under-determined. For all other geometrical settings, the observed differences were not such that one method could be straightforwardly identified as being in favor over all other methods. Especially in terms of noise, the behavior of the mean standard deviation and mean correlation coefficients was adverse. This means, for instance, the pixel standard deviation was always larger for Joseph's method, but this effect was systematically accompanied by lower correlations between the pixels. The pixels in the distance-driven method were much more correlated because the distance-driven method simply enforced more binding between the pixels. In summary, the basic metric results show that the choice of the model had very little influence on performance for the task, even though important differences could be noted visually in terms of pixel noise variance and correlation among pixels.

The quantitative evaluation using task-based assessment was set up to further investigate if and, if so, how the differences observed for the basic metrics effect image quality. For the mathematical observer studies, the task was to classify lesions with a fuzzy or sharp boundary. The results of the ideal observer study and the channelized Hotelling observer study showed that the choice of the forward projection model had very little influence on the performance of a task. All observed differences of the AUC were often very small (< 0.002) and always not statistically significant. The human observer study was based on a detection task and was carried out for the more challenging SPS geometries; except for G3-SPS and G7-SPS. In this respect again, the AUC performance of all methods was comparable and small, and observed differences were not statistically significant.

In summary, the following general conclusions can be drawn. Each forward projection model has its own characteristics in terms of basic metrics, especially in terms of noise properties. However, these differences totally balance out when performing a task-based assessment. This means, Joseph's method, the distance-driven method, and the bilinear method can be seen as equivalent in terms of image quality. On the other hand, when deciding which of these forward projection models should be used in an iterative reconstruction algorithm, the crucial criteria is driven by the user's goal and/or by the user's preferences. For instance, if the computational cost plays an important role, Joseph's method should be in favor of the other two methods.

CHAPTER 7

Challenges posed by Statistical Weights and Data Redundancies

This chapter describes the challenges posed by statistical weights and data redundancies in iterative CT image reconstruction. The solution to such statistical iterative reconstruction methods is often found by solving the maximum likelihood solution, particularly because the problem can be reduced to a weighted least squares problem. Given that a CT reconstruction problem is mildly ill-posed and, therefore, tends to be unstable, any kind of regularization is needed. There are two possible regularization methods: i) a regularization defined by stopping after a moderate number of iterates; and, ii) a regularization based on a penalty term. In this study we investigated the impact of both regularization options on image quality in a statistical image reconstruction problem.

Section 7.1 motivates the study and further shows the limitations of the study design. Next, in Section 7.2, the experimental setup is explained, including information about the geometrical settings, data simulation, phantom selection, the forward projection model, and the selection of the reconstruction algorithms. The concept of the parameter selection of one of the applied reconstruction algorithms, namely for the ICD method, is described in Section 7.3. This is followed by Section 7.4, which covers the first part of the study: the influence of a statistical weighting matrix. That section provides both the information about the noise statistics model and the results. Section 7.5 contains the second part of the study: the influence of a data redundancy weighting. As in Section 7.4, first the data-handling concept is described, followed by a representation of the results. Finally, a discussion and conclusion is given in Section 7.6.

Parts of this work have already been published in Schmitt et al. [Schm 13]. A follow-up study that investigated the influence of statistical weights together with a variety of edge preserving parameters on IQ may be found in Hahn et al. [Hahn 15b]. This work focused on a task-based assessment of IQ and has not been included in the dissertation.

7.1   Motivation and Limitations

Statistical iterative reconstruction methods are currently under extensive investigation in the CT community. Generally, there are many ways to formulate a statistical reconstruction method for X-ray CT. In particular, the maximum likelihood solution without and with constraints on the image appear highly popular. In both cases the solution can be reduced to a weighted least squares problem. The solution of a non-constrained iterative image reconstruction converges towards the maximum likelihood solution as described in Section 4.2.1. The final reconstruction needs to be defined as the application of a finite number of iteration steps since the CT reconstruction problem is mildly ill-posed. Then stopping is essential since the algorithm tends to be unstable [Vekl 87]. When following the approach of a constrained image reconstruction problem (Sec. 4.2.2), the regularization is not left to the iteration number; it is enforced directly by the constraint. Hence, the algorithm stops automatically when the minimum of the objective function is reached or when the change in the objective function becomes smaller than a pre-defined threshold.

Even though it is well-known that the iteration number creates a (shift-variant) trade-off between resolution and noise, the approach of stopping after a certain number of iterations has been found useful in nuclear medicine [Hebe 88, Llac 89], i.e., in positron emission tomography (PET) imaging. The question is whether this regularization approach is also effective in CT imaging. This was investigated by applying a statistical weighting matrix that covered essential aspects of CT imaging, specifically non-uniform statistical weights and data redundancies. Each of these two aspects was investigated separately and may be found below in Part I and Part II of the study.

As already mentioned in the previous chapter, it is essential for a meaningful observation to account for different noise realization and variations in geometry. Therefore, to mitigate this difficulty, the focus of the study was on 2D rather than on 3D reconstructions.

7.2   Experimental Setup

7.2.1   Study Setup

The study is subdivided into two parts. Each part corresponds to another interpretation of the statistical weighting matrix $\mathbf{W}$. In the first part, the statistical weighting matrix represents non-uniform weights. In the present work, the non-uniform weights are related to a bowtie filter in the beam path. Another popular example of the appearance of non-uniform weights in CT imaging is the use of tube current modulation (TCM); this, however, was not investigated in the present work. In the second part,

$\mathbf{W}$ takes the information about redundant data in the sinogram. A detailed mathematical description of each statistical weighting matrix is given in the corresponding study part in Section 7.4 and Section 7.5, respectively.

7.2.2 Geometrical Settings

The identical 16 geometrical (parametric) settings as defined in Section 6.3.2 were used in this study since this variety allows the probing of various features and thereby represents contemporary CT scan settings. The label concept and the parameter settings for each geometry are given on page 67.

Each part of that study considers another (sub-)set of the geometrical settings due to the specific design and interpretation of the statistical weighting matrix in each part. In the first part, the focus was on the full-scan geometrical settings, which in detail were G1, G2, G5, and G6, in moderate and small pixel size. In the second part, the focus was primarily on the short-scan geometries G3, G4, G7, and G8, again for both moderate and small pixel size. However, for reasons of comparability, in the second part the full-scan geometries have also been evaluated. This means, Part I considers 8 geometrical settings in total and Part II considers 16 geometrical settings.

7.2.3 Phantom Selection and Data Simulation

The simulations were carried out using the FORBILD head phantom and a uniform head phantom. Whereas the FORBILD head phantom was used in both parts of the study, the uniform head phantom was only used in Part I. The uniform head phantom consisted only of the outer two ellipses of the FORBILD head phantom with a background value of 50 HU as shown in Figure 7.2.

All data simulations were performed in fan-beam geometry for a $3^{\rm rd}$ generation CT scanner with a curved detector. In detail, the same concept of modeling the rays with the same modeling parameters as described in Section 6.3.3 were used in this study. This means, due to a sub-sampling of the focal spot, each detector element, each view, and each ray were modeled as an analytical, non-linear average of 405 line integrals. In total, 50 noise realizations for each geometrical setting and each phantom were available. All data simulation and image reconstruction parameters are listed in Table 6.1.

7.2.4 Reconstruction Methods

In this study, the preferences regarding the objective function were to: i) always include a constant diagonal weighting matrix $\mathbf{W}$; and, ii) optionally make use of a regularization term $\Phi_R(\boldsymbol{f})$. In the event that the regularization term was not equal to zero, it was always defined as convex. Both the Fair potential and the quadratic potential were used to define Φ_R (see Sec. 4.2.2).

The solution of the non-constrained image reconstruction problem ($\Phi_R = 0$) was found by the application of the Landweber algorithm (see Sec. 4.3.1). The Landweber algorithm tries to find the minimum-norm minimizer of a weighted least squares (WLS) problem as defined in Equation 4.8. The minimizer was, however, never reached since reconstruction was stopped after a moderate number of iterates. In total, 1000 iterations were performed and every fifth iterate was stored, starting with the result from the first iteration. All reconstructions were initialized with an image vector, $\boldsymbol{f}^{(0)}$, which was the zero vector. The converge controlling parameter η was chosen as 0.90 times $2/\sigma_{\mathrm{max}}$, where the maximum singular value of $\tilde{\mathbf{A}}$, where $\tilde{\mathbf{A}} = \mathbf{WA}$, was determined using five iterations of the power method. The reconstructed images below 40 iterations were neglected in the evaluations because resolution was too low for these images to be of interest.

The solution of the constrained image reconstruction problem ($\Phi_R \neq 0$) was found with the ICD algorithm (see Sec. 4.3.2). The algorithm was initialized with an image vector, $\boldsymbol{f}^{(0)}$, that was equal to the zero vector. The parameter ϵ in the Fair potential function was decided to be 50, and 500 for the quadratic potential. In the ICD algorithm, only the final image reconstruction was stored. We call a reconstruction the final image reconstruction when the change of each pixel value in the reconstructed volume from one to the next iteration became smaller than $10^{-5}/\mathrm{cm}$ (i.e., about 0.05 HU). The sharpness of the final image reconstruction is primarily determined by the regularization parameter, β_R. Therefore, a pre-study was required to elaborate the relationship of the final image resolution, $\nu_{0.5}$, and the regularization parameter β_R. The procedure of adapting the regularization parameter is described below in Section 7.3. In this work, the target resolution of all final reconstructions obtained with ICD was selected as 4.75/cm.

Sometimes, the reconstruction results with $\mathbf{W} = \mathbf{I}$ are desired when explicitly no statistical weighting matrix is applied, i.e., for a comparison of the results obtained with and without a statistical matrix. For a distinction, we introduce the following labeling concept. Hereafter, let "LS" refer to the least squares solution, i.e., $\mathbf{W} = \mathbf{I}$, and let "WLS" refer to the weighted least squares solution, i.e., $\mathbf{W} \neq \mathbf{I}$, both obtained under the application of the Landweber algorithm[1]. Reconstructions obtained from a penalized image reconstruction problem under the application of the ICD algorithm additionally receive the initial letter "P" plus the short name of the potential function that was used, where the short name "QP" refers to the quadratic potential and "FP" refers to the Fair potential. For instance, PWLS-QP denotes the result obtained from a penalized weighted least squares image reconstruction problem using the quadratic penalty.

7.2.5   Forward Projection Model

The forward projection matrix $\mathbf{A}$ was formed using the distance-driven method. That choice was related to the results obtained from the quantitative evaluation in the previous chapter, where Joseph's method and the bilinear method caused a relatively high bias in the short-scan geometries without FFS for the small pixel size compared to the distance-driven method.

---

[1]Our labeling concept intentionally neglects the fact that the Landweber algorithm was quasi-penalized by the application of a fixed number of iterates.

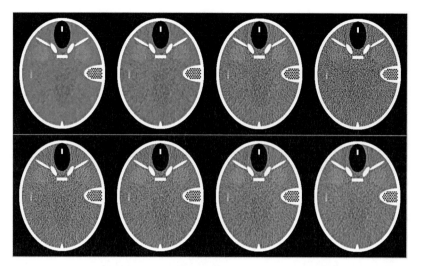

Figure 7.1: Reconstruction results of the FORBILD head phantom for the geometrical setting G1-MPS obtained with PWLS-FP under variation of the parameters ϵ and β_R. $\epsilon = 1, 5, 50, 500$ and $\beta_R = 0.4332$ *(first row, left to right)*. $\epsilon = 50$ and $\beta_R = 0.2, 0.4, 0.6, 0.8$ *(second row, left to right)*. Display window: c/w=50/200 HU.

7.3   Concept of ICD Parameter Selection

The ICD algorithm was applied to find the solution of the constrained image reconstruction problem. This means that, in comparison to the Landweber algorithm, the ICD algorithm has, due to the additional regularization term Φ_R in the objective function, two additional, freely selectable parameters, namely ϵ and β_R. Whereas the parameter ϵ mainly influences the final image impression, β_R almost exclusively controls the resolution in the final image reconstruction. Figure 7.1 displays these effects for the geometrical setting G1-MPS. All reconstructions in that image were obtained with the Fair potential (PWLS-FP) under variation of ϵ and β_R. In the first row, the reconstructions from left to right correspond to $\epsilon = 1, 5, 50, 500$, while the regularization parameter was always the same, $\beta_R = 0.4332$. In the second row, the reconstructions from left to right correspond to $\beta_R = 0.2, 0.4, 0.6, 0.8$, while the potential function parameter was always the same, $\epsilon = 50$. As clearly visible, the image impression of the reconstruction in the first row changes from left to right. For a small ϵ the image impression is more similar to a reconstruction obtained using a total variation constraint, whereas for large ϵ the image impression is more similar to a reconstruction obtained using FBP reconstruction. The reconstructions in the second row highlight the loss of resolution with an increasing β_R value.

For the constrained image reconstruction problem, the parameter ϵ in the potential function was chosen as 50 for the Fair potential and 500 for the quadratic potential. Both choices aim an FBP-like image impression.

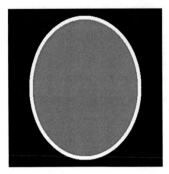

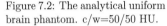

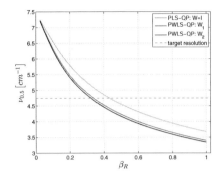

Figure 7.2: The analytical uniform brain phantom. c/w=50/50 HU.

Figure 7.3: Resolution, $\nu_{0.5}$, as a function of the regularization parameter β_R obtained with three different weighting matrices for G1-MPS.

While the Landweber algorithm yields a sequence of iterates, with each iterate corresponding to another spatial resolution, the reconstructed image obtained with the ICD method corresponds to exactly one resolution level. By having ϵ defined, the dependence of the final image resolution on β_R can now be investigated.

From the previous study, the range of possible resolution values obtained with the Landweber algorithm is already known. That was $\nu_{0.5} \in [1.55, 6.29]$/cm for the moderate pixel size (MPS) and $\nu_{0.5} \in [1.58, 5.43]$/cm for the small pixel size (SPS), when considering all available geometries. For SPS the highest resolution value was limited by the short-scan geometries without flying focal spot. With respect to the geometrical settings as given in Chapter 6, the ICD algorithm should aim for a resolution in the lowest common interval, which is a resolution between 1.58/cm and 5.43/cm. In this study, the target resolution was set to $\nu_{0.5} = 4.75$/cm, which is a moderate resolution for brain imaging.

In order to find the parameter value β_R that yields the target resolution, a series of reconstructions was evaluated all solving the objective function using the quadratic potential. Each reconstruction in the series corresponds to another β_R value. The regularization parameter was gradually increased by steps of 0.025, starting at $\beta_R = 0.025$ and stopping at $\beta_R = 1.0$, which yield in total 40 reconstructions. Then, for each reconstruction, the resolution was determined as described in Section 5.2.1. Finally, the value of β_R that yield $\nu_{0.5} = 4.75$/cm was found using linear interpolation. Note that by using the quadratic potential the approach of determining the resolution is still valid since the reconstruction process is linear.

That procedure was repeated for all statistical weighting matrices and all geometrical settings used throughout this chapter. Note that the elements of $\mathbf{W}$ were scaled so that the elements of $\mathbf{W}$ for the central ray had a mean value of one; this scaling enabled maintaining the same resolution at high contrast for all objective functions. Then, for the Fair and quadratic potential, the same values of β_R may be used.

Figure 7.3 represents typical resolution curves obtained with the approach described above for the geometrical setting G1-MPS for three different weighting matrices. The plot shows: i) that resolution is a smooth function depending on the regularization parameter; and, ii) that a statistical weighting matrix also affects the resolution curve. In the given plot, the target resolution is obtained for the following β_R values: $\beta_R = 0.4332$ for $\mathbf{W} = \mathbf{I}$, $\beta_R = 0.3396$ for $\mathbf{W}_1$, and $\beta_R = 0.3230$ for $\mathbf{W}_2$.

The results for β_R in dependence of the design of the statistical weighting matrix and the geometrical setting are given below in the corresponding study parts.

7.4   Part I: Reconstruction with Statistical Weights

X-ray beam filtration as described in Section 2.6 has become a gold standard in most of the commercially available CT scanners. Often, a flat filter or a small piecewise shaped material called bowtie filter (see Fig. 2.7), is placed between the X-ray tube and the patient with the primarily goal to reduce the effective dose imparted to the patient. Usually, multiple bowtie filters are provided in the CT scanner due to the quite different anatomy of the human body from head to heel. A measurement setup with a bowtie filter lets the number of photons vary with the fan angle γ. This aspect can be addressed in iterative reconstruction by including a statistical weighting matrix $\mathbf{W}$. The influence of such a non-uniform statistical weighting matrix is investigated in this present part of the study for two differently shaped bowtie filters.

7.4.1   Bowtie Filter Model

Let N_{in} be the number of photons that are emitted by the X-ray tube and let $g_B(\lambda)$ be a function that models the presence of the bowtie filter. Then, the variation of the number of photons entering the scanning object, N_{ph}, is defined by

$$N_{\mathrm{ph}}(\gamma) = N_{\mathrm{in}}\, e^{-g_B(\lambda)}. \tag{7.1}$$

To define $g_B(\lambda)$ we follow the approach described in Wunderlich et al. [Wund 08]. Let the bowtie filter be designed for a centered circular water cylinder of radius r_w with a linear attenuation function $\mu_w(\boldsymbol{x})$ and let $g_w(\lambda)$ be the line integral $\mathcal{L}(\lambda, \gamma)$ for a fixed angle λ through that water phantom. Furthermore, let d_B be the bowtie filter thickness at $\gamma = 0$ and let μ_B be the linear attenuation coefficient of the bowtie filter itself. Then, the bowtie filter modeling function may be expressed as

$$g_B(\lambda) = g_w(0) - g_w(\lambda) + d_B\,\mu_B. \tag{7.2}$$

Note that $g_B(\lambda)$ is circularly symmetric for a centered object. Thus, g_w is defined by any fixed value of λ. A schematic drawing of the X-ray filtration with the parameters given above is shown in Figure 7.4.

7.4.2   Experimental Details

7.4.2.1   Parameter Selection

The presence of the bowtie filter is introduced as part of a (non-uniform) statistical weighting matrix $\mathbf{W}$, where $\mathbf{W}$ is interpreted as the covariance matrix for the mea-

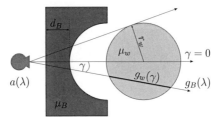

X-ray tube   bowtie filter   symmetric water cylinder

Figure 7.4: Schematic drawing for the definition of the bowtie filter modeling function $g_B(\lambda)$ that is designed for a centered circular water cylinder with radius r_w.

surements. Since all measurements are assumed to be statistically independent, $\mathbf{W}$ is a diagonal matrix. Each element on the diagonal represents the variance of one measurement. This variance is equal to the inverse of the number of photons reaching the detector. Thus, each statistical matrix element takes the value

$$W_{ij}^{-1} = N_{\mathrm{ph}}(\gamma)\, e^{-g_f(\lambda,\gamma)}, \qquad (7.3)$$

where $N_{\mathrm{ph}}(\gamma)$ are the remaining photons after the bowtie filter as defined in Equation 7.1.

Note that the variance is itself influenced by the attenuation property of the interrogated object, the number of incoming photons, and the shape of the bowtie filter. In total, two differently shaped bowtie filters were used. Both are shown in Figure 7.5 and are referred to as bowtie filter 1 and 2, respectively. The bowtie filters were assumed to be of aluminum with $r_w = 8$ cm and $d_B = 500$ μm. The simulation was based on a linear attenuation value of $\mu_B = 0.545$/cm which corresponds to an 80 keV X-ray photon spectra [Hubb 04]. The shape of bowtie filter 2 was designed such that the distribution of the photons after the filter was a smooth Gaussian-shaped curve whereas the curve of bowtie filter 1 is non-smooth. The normalized number of photons obtained after both bowtie filters is displayed in Figure 7.6.

In addition to these bowtie filter choices, a statistical weighting matrix that was equal to the identity matrix was applied. The reconstruction for $\mathbf{W} = \mathbf{I}$ corresponds to a (penalized) least squares solution, whereas W_{ij}^{-1} (Eq. 7.3) refers to a (penalized) weighted least squares solution. Hereafter, let $\mathbf{W}_1$ and $\mathbf{W}_2$ denote the weighting matrix obtained by considering bowtie filter 1 and 2, respectively.

To study the impact of such a non-uniform statistical weighting matrix, only the full-scan geometries in both pixels sizes were considered. This was largely because possible data redundancy effects should be eliminated.

For the application of the ICD algorithm, the values of β_R were adapted using the approach described in Section 7.3 to obtain a target resolution of $\nu_{0.5} = 4.75$/cm. Table 7.1 lists all values of β_R for each full scan setting (G1, G2, G5, and G6) in both pixel sizes (MPS and SPS) and for all weighting matrices ($\mathbf{W} = \mathbf{I}$, $\mathbf{W}_1$, and $\mathbf{W}_2$).

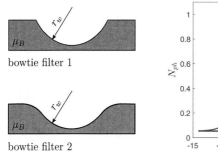

Figure 7.5: Two differently shaped bowtie filters were simulated.

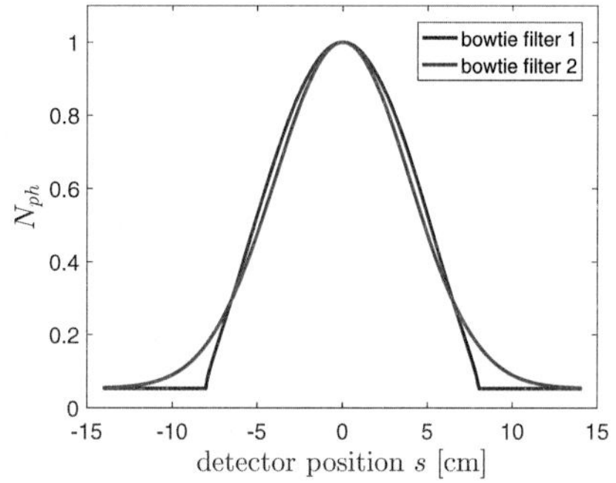

Figure 7.6: Normalized number of photons, N_{ph}, after bowtie filter 1 and 2, respectively.

	geometrical setting			
	G1	G2	G5	G6
MPS: PLS-QP (no bowtie filter $\to \mathbf{W} = \mathbf{I}$)	0.4332	0.8787	0.4101	0.8316
MPS: $\mathrm{PW_1LS}$-QP (bowtie filter 1 $\to \mathbf{W}_1$)	0.3396	0.6874	0.3222	0.6522
MPS: $\mathrm{PW_2LS}$-QP (bowtie filter 2 $\to \mathbf{W}_2$)	0.3230	0.6537	0.3070	0.6212
SPS: PLS-QP (no bowtie filter $\to \mathbf{W} = \mathbf{I}$)	0.3180	0.6471	0.2988	0.6105
SPS: $\mathrm{PW_1LS}$-QP (bowtie filter 1 $\to \mathbf{W}_1$)	0.2463	0.5007	0.2322	0.4738
SPS: $\mathrm{PW_2LS}$-QP (bowtie filter 2 $\to \mathbf{W}_2$)	0.2343	0.4759	0.2209	0.4508

Table 7.1: Part I: regularization parameter values β_R for the ICD algorithm for the different geometrical settings and weighting matrices.

Note that the given regularization parameters have been applied for both potential functions.

7.4.2.2   Image Quality Assessment

Image quality was assessed using basic metrics. Spatial resolution, mean absolute reconstruction error (bias), and noise was evaluated as described in detail in Sections 5.2.1, 5.2.4 and 5.2.5. Since other artifacts have been observed in some reconstructions (see below), the bias was evaluated as defined in Equation 5.3 in two additional ROIs. The first region of interest is a specifically defined centered ring-shaped area in the uniform brain phantom as displayed on the left in Figure 7.7. The outer radius of this region was 8.5 cm and the inner radius was 7.5 cm. The second region corresponds to an elliptical-shaped segment close to the right ear in the FORBILD head phantom as displayed on the right in Figure 7.7. To avoid confusion,

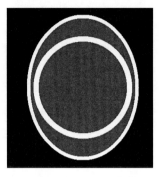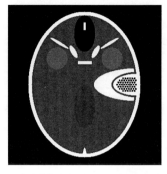

Figure 7.7: Representation of the two additional defined evaluation regions for the calculation of the mean reconstruction error called $\bar{\sigma}_r$ and $\bar{\sigma}_{ES}$, respectively. The evaluation region of $\bar{\sigma}_r$ corresponds to the highlighted white additional ring-shaped area in the uniform brain phantom *(left)*. The elliptical-shaped segment in the FOR-BILD head phantom was used for the calculation of $\bar{\sigma}_{ES}$ *(right)*. Display window: c/w=50/100 HU.

the bias corresponding to the full background of the phantom is called $\bar{\sigma}_b$, whereas the bias referred to in the first and second ROI is called $\bar{\sigma}_r$ and $\bar{\sigma}_{ES}$, respectively.

7.4.3   Results

This section provides the results obtained in the IQ assessment. First the results obtained with the Landweber algorithm (Sec. 7.4.3.1) followed by the results obtained with the ICD method (Sec. 7.4.3.2) are presented. Both sub-sections provide information about the display concept, some reconstruction examples, and the figures-of-merit.

7.4.3.1   Influence of Statistical Weights

Figure-of-Merit.   Since slight differences in spatial resolution were observed for each geometrical setting and for each representation of the weighting matrix, we display all figures-of-merit as a function of $\nu_{0.5}$. Linear interpolation was used to bring all results on the same set of samples of $\nu_{0.5}$.

Display concept.   Since the analysis generated a large number of curves, we follow the same concept as in Section 6.5 of reducing the dependence over all geometries to one summary plot. The summary plot is the metric as a function of $\nu_{0.5}$ as obtained after averaging over all four full scan geometries, depending on the pixel size and the value of $\nu_{0.5}$. Figure 7.8 shows for each realization of the weighting matrix the geometry specific result of the mean reconstruction error $\bar{\sigma}_b$ together with the average over the four geometries.

Note that the frequency range for MPS and SPS was slightly different, namely $\nu_{0.5} \in [2.26, 6.72]/\text{cm}$ for MPS and $\nu_{0.5} \in [2.31, 7.45]/\text{cm}$ for SPS. Results corre-

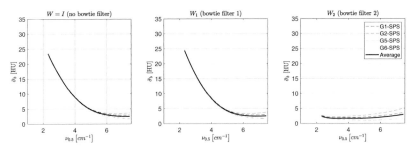

Figure 7.8: Mean and averaged absolute reconstruction error $\bar{\sigma}_b$ as a function of $\nu_{0.5}$ for all four full scan geometries in SPS. To minimize the number of dependencies the averaged metric (black bold line) is used in the summary plot.

sponding to different metrics or pixel sizes are shown in separate figures, with the abbreviation MPS or SPS to emphasize to which pixel size a figure or plot corresponds. Three different tones of gray represent the full scan results: light gray for "LS" (no bowtie filter), medium gray for "PW_1LS" (bowtie filter 1), and dark gray for "PW_2LS" (bowtie filter 2).

Visual appearance of some reconstructions.   Figure 7.9 displays noiseless reconstructions after 251 Landweber iterations. The figure shows from top to bottom the full scan reconstructions for the geometrical settings G1-MPS, G2-MPS, G5-MPS, and G6-MPS obtained using the following weighting matrices (first to third column): $W = I$ (no bowtie filter), W_1 (bowtie filter 1), W_2 (bowtie filter 2). Figure 7.11 shows the difference images LS – W_1LS, LS – W_2LS, and W_1LS – W_2LS obtained from the noiseless reconstruction of G1-MPS after 251 iterations. Figure 7.10 displays noisy reconstructions for MPS in the same arrangement as in Figure 7.9. Note that the grayscale window is more compressed for the noiseless reconstructions to emphasize the differences in image quality between the reconstructions. We only provide the reconstructions for MPS since the SPS reconstructions do not convey more information.

The influence of a statistical weighting matrix on image quality is best visible in all noiseless image reconstructions in terms of strong artifacts. For example, the use of bowtie filter 1 causes strong streak artifacts especially close to the right ear and in the upper part of the FORBILD head phantom, as well as a prominent ring-shaped artifact. That ring-shaped artifact can be traced back to the fact that the function $N_{ph}(\gamma)$ is non-smooth since the radius of that ring has the same radius as the one of bowtie filter 1, namely $r_w = 8$ cm. By iterating a very long time, the thickness of the ring decreases accordingly. Finally, the ring disappears completely first after 1586 iterates for G1-MPS, after 1721 iterates for G2-MPS, after 1636 iterates for G5-MPS, and after 1761 iterates for G6-MPS. Figure 7.12 shows the reconstructed images of the uniform brain phantom after 251 and after 1586 iterations for G1-MPS together with the corresponding vertical profile through the origin of the phantom. For SPS,

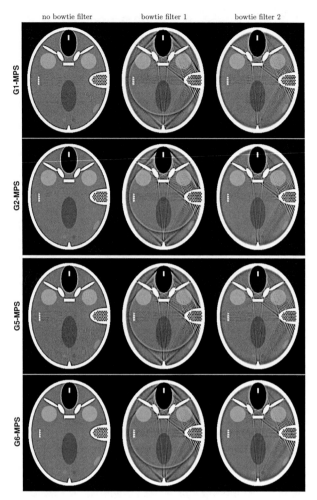

Figure 7.9: Noiseless reconstructions obtained with the Landweber algorithm after 251 iterations for the full scan geometries G1-MPS, G2-MPS, G5-MPS, and G6-MPS *(first to fourth row)* with $\mathbf{W} = \mathbf{I}$ (no bowtie filter), $\mathbf{W}_1$ (bowtie filter 1), and $\mathbf{W}_2$ (bowtie filter 2) *(first to third column)*. Display window: c/w=50/40 HU.

the iteration number that is needed to make the ring disappear is much higher than for MPS, namely 2436 for G1, 2871 for G2, 2671 for G5, and 2896 for G6.

The strong streak artifacts can be traced back to a lack of smoothness in the statistical weights since they are only present in more complex phantoms. Reconstructions of the uniform brain phantom (see Fig. 7.12) do not show streaks for any

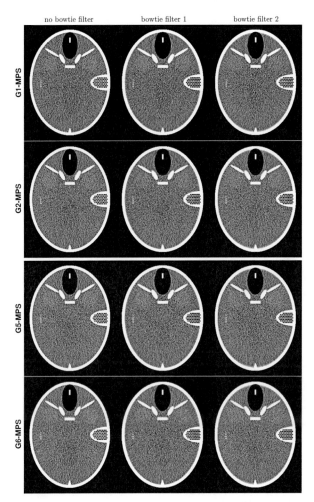

Figure 7.10: Noisy reconstructions obtained with the Landweber algorithm after 251 iterations for the full scan geometries G1-MPS, G2-MPS, G5-MPS, and G6-MPS *(first to fourth row)* with $\mathbf{W} = \mathbf{I}$ (no bowtie filter), $\mathbf{W}_1$ (bowtie filter 1), and $\mathbf{W}_2$ (bowtie filter 2) *(first to third column)*. Display window: c/w=50/200 HU.

of the bowtie filters whereas the streaks are present for both bowtie filter options in the reconstructions of the FORBILD head phantom.

The noisy reconstructions show that some of the artifacts are superimposed by the noise. However, most of them, especially the ring artifact and the streaks close to the right ear, are still recognizable at least after 251 iterations.

a) G1-MPS: LS-W$_1$LS   b) G1-MPS: LS-W$_2$LS   c) G1-MPS: W$_1$LS-W$_2$LS

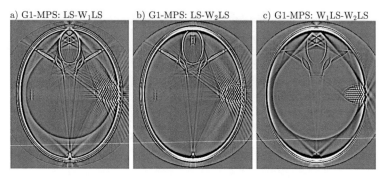

Figure 7.11: Difference images of the noiseless reconstruction for G1-MPS after 251 Landweber iterations: *a)* LS – W$_1$LS, *b)* LS – W$_2$LS, and *c)* W$_1$LS – W$_2$LS. Display window: c/w=30/60 HU.

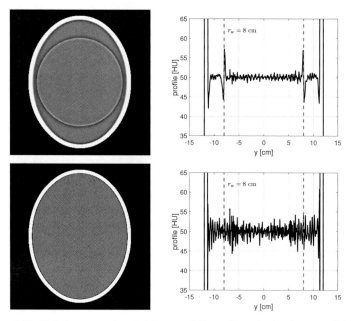

Figure 7.12: Noiseless reconstructed images of the uniform brain phantom *(left)* obtained with W$_1$LS after 251 *(top)* and 1586 iterations *(bottom)* for the geometrical setting G1-MPS. Display window: c/w = 50/100 HU. Vertical profiles through the origin of the phantom of the reconstructed images displayed on the left *(right)*.

	N_{iter}	$\bar{\sigma}_b$ [HU]	$\bar{\sigma}_{ES}$ [HU]	$\bar{\sigma}_r$ [HU]	$\bar{\sigma}_n$ [HU]	$\bar{\rho}_{45}$ [HU]	$\bar{\rho}_y$ [HU]
MPS: LS	256	1.32	1.37	0.62	44.40	−0.06	0.21
MPS: W_1LS	166	3.71	7.29	6.41	30.31	−0.02	0.18
MPS: W_2LS	166	2.58	6.76	0.60	32.61	−0.03	0.17
SPS: LS	231	2.04	1.70	0.25	49.36	0.29	0.61
SPS: W_1LS	151	5.68	7.47	7.16	33.42	0.35	0.58
SPS: W_2LS	151	5.15	6.90	0.60	35.18	0.33	0.58

Table 7.2: Quantitative metric results for G1-MPS and G1-SPS at $\nu_{0.5} = 4.75/\text{cm}$. The parameter N_{iter} denotes the number of iterations needed to reach the target resolution.

Reconstruction error.   Figure 7.13 compares the weighting matrices in terms of reconstruction error measured within: i) the full background of the FORBILD head phantom (first row); ii) the elliptical-shaped ROI within the FORBILD head phantom (second row); and, iii) the ring-shaped area in the uniform brain phantom (third row). Table 7.2 lists the results for the bias measures obtained at the target resolution of $\nu_{0.5} = 4.75/\text{cm}$ for the geometrical setting G1 for both pixel sizes.

Including a statistical weighting matrix with $\mathbf{W} \neq \mathbf{I}$ increased the nominal values of all error measures significantly. While the bias for the LS solutions remains largely stable over the frequency range $\nu_{0.5}$, the bias for bowtie filter 1 and 2 starts with a high value at low frequency and then slowly decrease when going to higher frequencies. In example, for MPS the bias σ_b was between 0.89 HU and 2.36 HU for LS, between 1.97 HU and 16.35 HU for W_1LS, and between 1.81 HU and 16.84 HU for W_2LS. For SPS the bias σ_b was between 1.66 HU and 2.92 HU for LS, between 2.51 HU and 23.41 HU for W_1LS, and between 2.45 HU and 24.34 HU for W_2LS. Note that the $\bar{\sigma}_b$ values measured in the LS reconstructions are nearly in the whole frequency range below the weighted least squares solutions.

The same strong observation is valid for the bias metric $\bar{\sigma}_{ES}$ with the only exception that the nominal values of the weighted least squares solutions stay always above the LS solution in the whole frequency range. The ranking from high to low bias values was W_1LS, W_2LS, and LS. For both MPS and SPS, the curve trends were very similar and even within the same value range.

The bias metric $\bar{\sigma}_r$ measures the artifacts within the ring-shaped area in the uniform brain phantom. Since no ring artifact is observed in the LS and W_2LS reconstructions, the bias for these two curves is much lower than for W_1LS. The plots also show that the $\bar{\sigma}_r$ for W_1LS slowly decrease, which means many iterations are needed to reach the nominal values of the other two curves.

Apart from a significant increase of artifacts in the image, a statistical weighting matrix helps to speed up convergence in terms of necessary iterations, called N_{iter}, that are needed to reach a certain resolution (see Tab. 7.2). For example, the resolution $\nu_{0.5} = 4.75/\text{cm}$ was reached 90 iterations earlier for MPS and 80 iterations

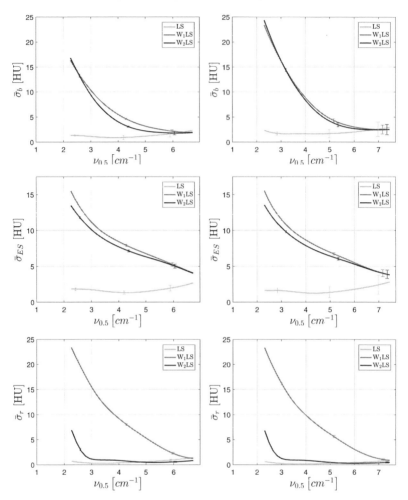

Figure 7.13: Geometry-averaged reconstruction error measures for MPS *(left)* and SPS *(right)* evaluated using the full background of the FORBILD head phantom *(first row)*, the elliptical-shaped segment *(second row)*, and the ring-shaped area in the uniform brain phantom *(third row)* as shown in Figure 7.7.

earlier for SPS when applying bowtie filter 1 and 2, respectively. Nevertheless, that advantage comes with a heavy increase in bias due to non-smooth statistical weights.

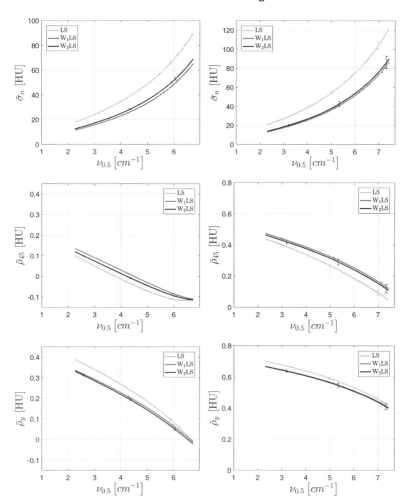

Figure 7.14: Geometry-averaged noise metrics for MPS *(left)* and SPS *(right)*. Mean standard deviation *(first row)*, correlation coefficient at 45 degrees, *(second row)* and correlation coefficient in y *(third row)*.

Noise.   Figure 7.14 compares the statistical weighting matrices in terms of noise performance for MPS and SPS, respectively. The results for the correlation coefficient in x are not shown because they did not convey any additional information.

For both MPS and SPS, the dependence of noise metrics on frequency and geometry was very similar for all objective functions. Including a statistical weighting matrix with $\mathbf{W} \neq \mathbf{I}$ helps to reduce the mean standard deviation by up to 24.8 HU for

MPS and by 35.2 HU for SPS, respectively. Note that $\mathbf{W}_1$ always performed better than $\mathbf{W}_2$, specifically by up to 3.8 HU for MPS and by up to 3.2 HU for SPS.

The differences observed in the correlation coefficient were also significant. Using no weighting matrix ($\mathbf{W} = \mathbf{I}$) always yields the lowest correlation for $\bar{\rho}_{45}$ and always the highest correlation for $\bar{\rho}_y$. A statistical data weighting matrix shifted the correlation coefficients $\bar{\rho}_{45}$ by up to 0.04 for MPS and by up to 0.07 for SPS to higher values. The correlation coefficients $\bar{\rho}_y$ were shifted by up to 0.06 for MPS and by up to 0.03 for SPS to lower values. Thus, the correlation coefficients $\bar{\rho}_{45}$ and $\bar{\rho}_y$ are brought closer together.

7.4.3.2 Influence of Statistical Weights and Penalty Term

Figure-of-merit. All figures-of-merit represent the results obtained at the target resolution of $\nu_{0.5} = 4.75/\mathrm{cm}$.

Display concept. All results are displayed using a bar plot that shows the geometry specific result. Results corresponding to different metrics or pixel sizes are shown in separate figures, with the abbreviation MPS or SPS to emphasize to which pixel size a figure or plot corresponds. Each plot contains the result for the three different groups of objective functions referred to as "(W_i)LS", "P(W_i)LS-QP" and "P(W_i)LS-FP", where (W_i) indicates the absence or presence of a specific weighting matrix, called W_i with $i = 1, 2$. Three different tones of both contour and filling represent the different groups of objective functions: i) colored bars (contour and filling) are employed for the (weighted) least squares results; ii) gray colored bars (contour and filling) are employed for the penalized (weighted) least squares results using the quadratic potential; and, iii) transparent bars with gray colored contours are employed for the penalized (weighted) least squares results using the Fair potential.

Visual appearance of some reconstructions. Figure 7.15 displays from top to bottom noiseless reconstructions of the geometrical setting G1-MPS obtained from solving a (weighted) least squares problem, a penalized (weighted) least squares problem with quadratic penalty and a penalized (weighted) least squares problem with the Fair potential. The reconstructions in column one to three involve a statistical weighting matrix defined using no bowtie filter, bowtie filter 1, and bowtie filter 2. Figure 7.16 displays noisy reconstructions for MPS in the same arrangement as in Figure 7.15. Note that the grayscale window is more compressed for the noiseless reconstructions to emphasize the differences in image quality between the reconstructions. We only provide the reconstructions for one geometrical setting since the other reconstructions do not convey more information.

The noiseless reconstructions obtained from the three objective functions show that the use of a statistical weighting matrix has a strong impact on image quality for all three objective functions. Additional streak artifacts are present in all reconstructions where $\mathbf{W} \neq \mathbf{I}$. They can be found especially close to the right ear and close to the small bones in the upper part of the FORBILD head phantom. These artifacts were not present for the non-weighted (penalized) least squares solutions. Note that the streaks are less significant for the penalized weighted least squares

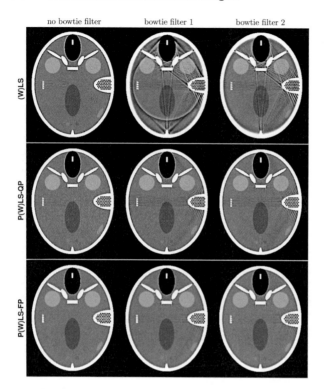

Figure 7.15:   Noiseless reconstructions of the geometrical setting G1-MPS obtained from solving a (weighted) least squares problem *(first row)*, a penalized (weighted) least squares problem with quadratic penalty *(second row)*, and a penalized (weighted) least squares problem with the Fair potential *(third row)* under application from $\mathbf{W} = \mathbf{I}$ (no bowtie filter), $\mathbf{W}_1$ (bowtie filter 1), and $\mathbf{W}_2$ (bowtie filter 2) *(first to third column)*. Display window: c/w=50/40 HU.

solutions where the penalty term helps to smooth them out. Also, the strong ring artifact that was observed in the $\mathbf{W}_1$LS solutions does not appear in the PW$_1$LS and PW$_2$LS reconstructions. This means, the penalized weighted least squares solution can deal much better with a non-smooth weighting matrix.

In the noisy reconstructions, most of the artifacts in the penalized (weighted) least squares solutions are superimposed by the noise and are, therefore, not so clearly visible. However, this is not the case for the noisy reconstructions obtained with the first objective function where most of the artifacts are still recognizable.

Reconstruction error.   Figure 7.17 compares the statistical weighting matrices and objective functions for both pixel sizes in terms of reconstruction error mea-

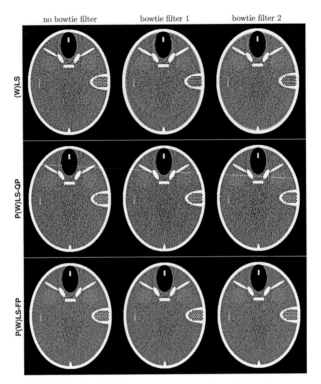

Figure 7.16: Noisy reconstructions of the geometrical setting G1-MPS obtained from solving a (weighted) least squares problem *(first row)* , a penalized (weighted) least squares problem with quadratic penalty *(second row)*, and a penalized (weighted) least squares problem with the Fair potential *(third row)* under application from $\mathbf{W} = \mathbf{I}$ (no bowtie filter), $\mathbf{W}_1$ (bowtie filter 1), and $\mathbf{W}_2$ (bowtie filter 2) *(first to third column)*. Display window: c/w=50/200 HU.

sured within: i) the full background of the FORBILD head phantom; ii) within the elliptical-shaped ROI within the FORBILD head phantom; and, iii) within the ring-shaped area in the uniform brain phantom, referred to as $\bar{\sigma}_b$, $\bar{\sigma}_{ES}$, and $\bar{\sigma}_r$.

The objective function has a global impact on the nominal values obtained for the reconstruction error for both pixel sizes, whereas the trends across the three groups of objective functions remain largely the same. For example, the overall bias is reduced when a penalty term is applied compared to the solution without a penalty, with the biggest effect observed with the edge-preserving potential. However, as already observed in the previous sections, a noticeable increase in bias is caused by including a statistical weighting matrix with $\mathbf{W} \neq \mathbf{I}$. That effect is observed also for the penalized weighted least squares solutions although with less extent. The measures

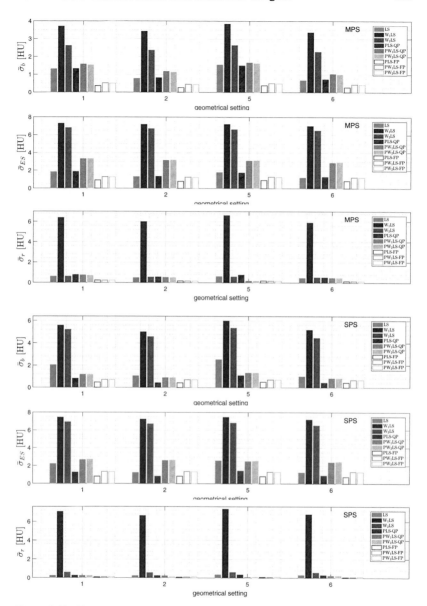

Figure 7.17: Reconstruction error measures $\bar{\sigma}_b$, $\bar{\sigma}_{ES}$, and $\bar{\sigma}_r$ for the three groups of objective functions including different statistical weighting matrices for MPS *(first and third row)* and SPS *(fourth and sixth row)*.

$\bar{\sigma}_b$ and $\bar{\sigma}_{ES}$ highlight best the extend of the streak artifacts. For example, the bias $\bar{\sigma}_b$ over all (W)LS results was between 0.7 HU and 3.8 HU for MPS, and between 1.0 HU and 6.0 HU for SPS. By comparison, these values dropped for the P(W)LS results and were in the range between 0.3 HU and 1.8 HU for MPS, and between 0.4 HU and 1.3 HU for SPS. Since no ring artifact is present for the penalized weighted least square solutions, the nominal values are very low, i.e., between 0.14 HU and 0.77 HU for MPS, and between 0.06 HU and 0.34 HU for SPS.

Noise. Figure 7.18 compares the statistical weighting matrices and objective functions in terms of noise performance for MPS and SPS, respectively. The results for the correlation coefficient in x are not shown because they did not convey any additional information.

As already indicated in the previous section, the objective function has a global impact on the nominal values obtained for each noise metric. Note that the effect of a statistical weighting matrix depends mostly on the underlying objective function, i.e., in some noise measures the nominal values increase, whereas in others the values decrease when a statistical weighting matrix is applied. However, the biggest changes in the nominal values of the noise metric were again observed for the edge-preserving potential.

A non-uniform statistical weighting matrix reduces the value of the mean standard deviation, $\bar{\sigma}_n$, in the WLS solution, whereas it is increased by up to 1.7 HU for MPS and by up to 1.9 HU for SPS in the PWLS solution compared to the PLS solution. Despite the small increase of the mean standard deviation, the absolute value of $\bar{\sigma}_n$ is much lower for the P(W)LS-FP reconstructions than for the (W)LS reconstructions, i.e., for MPS, $\bar{\sigma}_n$ was between 29.6 HU and 46.0 HU for (W)LS and between 17.0 HU and 19.2 HU for P(W)LS-FP, and, for SPS, $\bar{\sigma}_n$ was between 32.4 HU and 51.1 HU for (W)LS and between 16.3 HU and 18.1 HU for P(W)LS-FP. The correlation between pixels is, in most cases, a little reduced compared to the solution without a statistical weighting matrix. This means that, the binding of the pixels is reduced, which in turn could cause the small increase in the mean standard deviation value.

7.4.4   Summary Discussion

In this part of the study, we have presented the influence of a (non-uniform) statistical weighting matrix on image quality. The investigation was divided into two subsections. In the first subsection, the focus was on the influence of a statistical weighting matrix on its own, which was investigated using images created through Landweber iterations, directly enabling the observation of the effects caused by the statistical weights. For the second subsection, a penalty term was added in the objective function. In total, two different statistical weighting matrices were applied which reflect bowtie filters of different shapes, specifically a non-smooth and a smooth bowtie filter. For reasons of comparison, a corresponding non-weighted solution was always evaluated. The subsections revealed fundamental differences in terms of image quality.

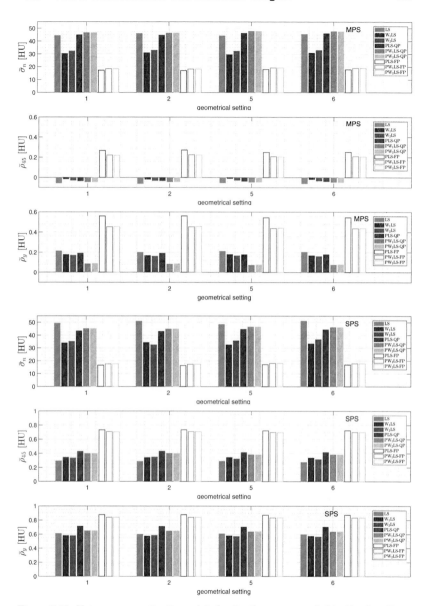

Figure 7.18: Noise measures $\bar{\sigma}_n$, $\bar{\rho}_{45}$ and $\bar{\rho}_y$ for the three groups of objective functions, including different statistical weighting matrices for MPS *(first and third row)* and SPS *(fourth and sixth row)*.

The results show that the presence of a statistical weighting matrix may easily introduce significant errors that differ from resolution errors. The salient observations were as follows:

- The WLS solution is very sensitive to a lack of smoothness in the statistical weighting matrix. Such a lack of smoothness can be caused by both a bowtie modeling function that is not smooth and by a strong change in attenuation, which mostly depends on the complexity of the irradiated object. Both effects may cause strong errors that may become visible as ring artifacts and/or as streak artifacts.

- The PWLS solutions are more stable than the WLS solution. The penalty term helps to remove most of the artifacts, i.e., the artifact that may be caused by a non-smooth bowtie modeling function. Even though a penalty term helps to improve the image quality significantly compared to a PLS solution, some of the artifacts remain in the reconstructed image. As the statistical weights have an enormous global impact on the nominal value of the bias at a fixed penalty term, the impact on the noise metrics is quite small and less predictable. For example, the mean standard deviation was reduced by up to 45% including statistical weights in the WLS solutions, whereas the opposite turned out for the PWLS solutions even though the values were very small ($< 11\%$).

In summary, it is not straightforward to state that the use of statistical weights which ideally model the physics of each measurement helps to improve the image reconstruction. It first depends on the design of the statistical weights. The smoother the statistical weighting matrix the better. However, redesigning of apparently realistic weights can become very challenging. Another important impact factor is the regularization approach. For instance, stopping after a given number of iterations, i.e., after 250 Landweber iterations, may be deemed satisfactory in terms of resolution; however, the user may generally observe that the image quality is not. In such situations, iterations far beyond the desired resolution are needed to first remove all reconstruction errors caused by the statistical weights and then post-smooth the result to attain the desired resolution. Under such circumstances, the penalized maximum-likelihood solution might be perceived as a more attractive reconstruction procedure. An alternative approach might be to initialize the reconstruction process with a filtered-backprojection procedure. However, in this case, it is important to understand what component of this first image remains when the reconstruction is completed.

7.5   Part II: Reconstruction with Data Redundancies

CT scans often involve redundant data, i.e., in a full scan each line integral is measured twice. However, there are situations where not all line integrals are always measured the same number of times. In particular, when performing a short scan, some line integrals are measured twice, whereas others are only measured once. This aspect needs to be carefully addressed in analytical reconstruction to avoid artifacts, whereas in iterative reconstruction there is a priori no need to use a careful data redundancy-handling concept. However, if desired such a concept can be introduced as a part of the definition of the weighting matrix $\mathbf{W}$. The influence of that matrix is investigated in this part of the study.

7.5.1 Data Redundancy Handling

A common approach to dealing with data redundancies in analytical image reconstruction was introduced by Parker in 1982 [Park 82]. By introducing a smooth weighting function, Parker showed that the image quality from a data set measured over 360° is essentially equivalent to that obtained from a minimal complete data set over 180°. In the present work, we used a Parker-like weight that was introduced by Noo et al. [Noo 02].

Let $c(\lambda)$ be a smooth function that is defined between the start and end point of the scan, called λ_s and λ_e, respectively. Further, let d be an angular interval over which $c(\lambda)$, defined as

$$
c(\lambda) = \begin{cases} \cos^2 \left(\frac{\pi(\lambda - \lambda_s - d)}{2d} \right) & \text{if } \lambda_s \leq \lambda < \lambda_s + d \\ 1 & \text{if } \lambda_s + d \leq \lambda < \lambda_e - d \,, \\ \cos^2 \left(\frac{\pi(\lambda - \lambda_e + d)}{2d} \right) & \text{if } \lambda_e - d \leq \lambda < \lambda_e \end{cases} \tag{7.4}
$$

drops smoothly from its maximum value of 1 to zero. The last equation is then used to define the data redundancy handling function, $m(\lambda, \gamma)$, that is given as

$$
m(\lambda, \gamma) = \frac{c(\lambda)}{c(\lambda) + c(\lambda + \pi - 2\gamma)} \,. \tag{7.5}
$$

Note, if d is small, the weighting function $m(\lambda, \gamma)$ is similar to that given in [Chen 06]. On the other hand, if d is large, $m(\lambda, \gamma)$ is similar to the Parker weight [Park 82].

7.5.2 Experimental Details

7.5.2.1 Parameter Selection

The data redundancy handling concept is introduced as part of the weighting matrix by setting the elements, $\mathbf{W}_{ij}^{-1}$, equal to the result of Equation 7.5. In total, three different angular intervals for d in the function $c(\lambda)$ have been chosen, which are $d = 0°, 5°, 30°$. Note that $d = 0°$ means that no data redundancy weight is applied since $m(\lambda, \gamma) = 1$. Thus, for $d = 0°$, the weighting matrix is represented by the identity matrix. This means that the solution for $d = 0°$ corresponds to a (penalized) least squares solution, whereas the settings $d \neq 0°$ refers to a (penalized) weighted least squares solution. Hereafter, let $\mathbf{W}_1$ and $\mathbf{W}_2$ denote the weighting matrix obtained by setting $d = 5°$ and $d = 30°$, respectively.

To study the impact of a Parker-like weight, both short scans and full scans were considered for both pixel sizes. In the full scan reconstructions, the weighting matrix was defined as the identity matrix.

	geometrical setting			
	G3	G4	G7	G8
MPS: PLS-QP ($d = 0° \rightarrow \mathbf{W} = \mathbf{I}$)	0.2313	0.4767	0.2170	0.4770
MPS: PW$_1$LS-QP ($d = 5° \rightarrow \mathbf{W}_1$)	0.2195	0.4525	0.2068	0.4264
MPS: PW$_2$LS-QP ($d = 30° \rightarrow \mathbf{W}_2$)	0.2181	0.4480	0.2055	0.4225
SPS: PLS-QP ($d = 0° \rightarrow \mathbf{W} = \mathbf{I}$)	0.1639	0.3505	0.1514	0.3265
SPS: PW$_1$LS-QP ($d = 5° \rightarrow \mathbf{W}_1$)	0.1550	0.3327	0.1442	0.3115
SPS: PW$_2$LS-QP ($d = 30° \rightarrow \mathbf{W}_2$)	0.1533	0.3292	0.1428	0.3085

Table 7.3: Part II: regularization parameter values β_R for the ICD algorithm for the different geometrical settings and weighting matrices.

For the application of the ICD algorithm, the values of β_R have been adapted using the approach described in Section 7.3 to obtain the target resolution of $\nu_{0.5} = 4.75/\text{cm}$. Table 7.3 lists all values of β_R for each short scan setting (G3, G4, G7, and G8) in both pixel sizes (MPS and SPS) and for all weighting matrices ($\mathbf{W} = \mathbf{I}$, $\mathbf{W}_1$, and $\mathbf{W}_2$). The regularization parameters for the full scan reconstructions for $\mathbf{W} = \mathbf{I}$ can be found in Table 7.1. Note that the given regularization parameters have been applied for both potential functions.

7.5.2.2 Image Quality Assessment

Image quality was assessed using basic metrics. In detail, spatial resolution and noise was evaluated as described in Sections 5.2.1 and 5.2.5, respectively. Furthermore, the mean absolute reconstruction error (bias) was analyzed as defined in Section 5.2.4. Since close to the right ear high reconstruction errors are expected, the bias was additionally calculated in a specifically defined region as displayed on the right side in Figure 7.7. The definition of calculating the bias for the evaluation regions is defined in Equation 5.3. To avoid confusion and to be consistent with the previous section, the bias corresponding to the highlighted region in Figure 7.7 is called $\bar{\sigma}_{ES}$, whereas the bias of the full background of the phantom is called $\bar{\sigma}_b$.

7.5.3   Results

This section provides the results obtained in the IQ assessment. The results obtained with the Landweber algorithm (Sec. 7.5.3.1) are first presented followed by the results obtained with the ICD method (Sec. 7.5.3.2). Both sub-sections provide information about the display concept, some reconstruction examples, and the figures-of-merit.

7.5.3.1 Influence of Statistical Weights

Figure-of-Merit. Since slight differences in spatial resolution were observed for each geometrical setting and for each representation of the weighting matrix, we

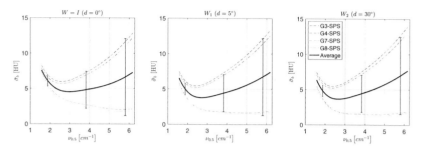

Figure 7.19: Mean and averaged absolute reconstruction error as a function of $\nu_{0.5}$ for all four short scan geometries in SPS. To minimize the number of dependencies, the averaged metric (black bold line) is used in the first summary plot.

display all figures-of-merit as a function of $\nu_{0.5}$. Linear interpolation was used to bring all results on the same set of samples of $\nu_{0.5}$.

Display concept.   Since the analysis generated a large number of curves, we follow the same concept as in Section 6.5 of reducing the dependency over all geometries to two summary plots. Results corresponding to different metrics or pixel sizes are shown in separate figures, with the abbreviation MPS or SPS to emphasize to which pixel size a figure or plot corresponds.

The first summary plot is the metric as a function of $\nu_{0.5}$ as obtained after averaging over all four short scan geometries, depending on the pixel size and the value of $\nu_{0.5}$. Figure 7.19 shows the geometry specific result of the mean reconstruction error $\bar{\sigma}_b$ together with the average over the four geometries for each realization of the weighting matrix. Note that the frequency range for MPS and SPS was slightly different, namely $\nu_{0.5} \in [1.55, 6.29]$/cm for MPS and $\nu_{0.5} \in [1.57, 6.17]$/cm for SPS. The second summary plot is a bar plot that shows the geometry specific average of the metric over the full $\nu_{0.5}$ interval. The second summary plot is only shown for the bias metrics since the curve divergence between the geometries with and without FFS is very prominent as shown in Figure 7.19.

Each summary plot contains the results of the short scan (SSC) and the full scan (FSC) geometry. Three different tones of gray represent the short scan results: light gray for "LS" ($d = 0°$), medium gray for "PW$_1$LS" ($d = 5°$), and dark gray for "PW$_2$LS" ($d = 30°$). The corresponding full scan result is always displayed as dashed black line and is labeled as "FSC: LS".

Visual appearance of some reconstructions.   Figure 7.20 displays noiseless reconstructions after 251 Landweber iterations. On the left side, the figure shows the short scan reconstructions for the geometrical settings G3-MPS, G4-MPS, G7-MPS, and G8-MPS obtained using a weighting matrix defined by $d = 0°, 5°, 30°$. On the right side, the corresponding full scan geometry reconstruction for the geometrical setting G1-MPS, G2-MPS, G5-MPS, and G6-MPS with a weighting matrix defined using $d = 0°$ are shown. Figure 7.21 shows the difference images LS − W$_1$LS, LS −

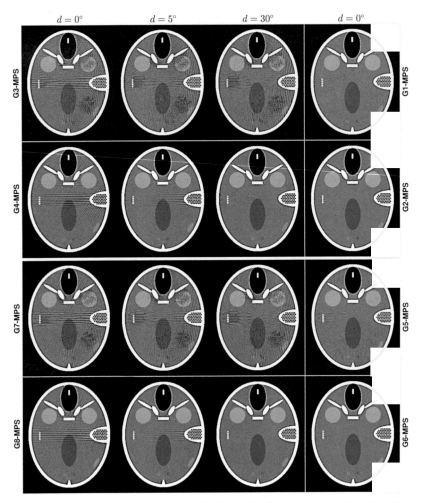

Figure 7.20: Noiseless reconstructions obtained with the Landweber algorithm after 251 iterations for the short scan geometries G3-MPS, G4-MPS, G7-MPS, and G8-MPS *(first to fourth row)* with $\mathbf{W} = \mathbf{I}$ $(d = 0°)$, $\mathbf{W}_1$ $(d = 5°)$, and $\mathbf{W}_2$ $(d = 30°)$ *(first to third column)*. Noiseless reconstructions of the corresponding full-scan geometry using $\mathbf{W} = \mathbf{I}$ $(d = 0°)$ *(fourth column)*; G1-MPS, G2-MPS, G5-MPS, and G6-MPS *(first to fourth row)*. Display window: c/w=50/40 HU.

a) G3-MPS: LS-W$_1$LS   b) G3-MPS: LS-W$_2$LS   c) G3-MPS: W$_1$LS-W$_2$LS

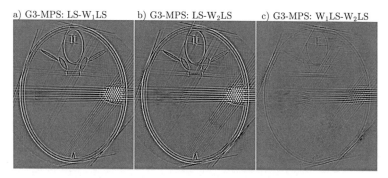

Figure 7.21: Difference images for the noiseless reconstruction for G3-MPS after 251 Landweber iterations: *a)* LS – W$_1$LS, *b)* LS – W$_2$LS, and *c)* W$_1$LS – W$_2$LS. Display window: c/w=30/60 HU.

W$_2$LS, and W$_1$LS – W$_2$LS obtained from the noiseless reconstructions of G3-MPS after 251 Landweber iterations. Figure 7.22 displays noisy reconstructions for MPS in the same arrangement as in figure 7.20. Note that the grayscale window is more compressed for the noiseless reconstructions to emphasize the differences in image quality between the reconstructions. We only provide the reconstructions for MPS since the SPS reconstructions do not convey more information.

As clearly recognizable in the noiseless reconstructions, the image quality is significantly improved by introducing a data redundancy weight as part of the weighting matrix. That effect is particularly more pronounced for the short scan reconstructions without FFS, i.e., for G3-MPS and G7-MPS, respectively. The artifacts that appear close to the right ear or close to the slanted bones in the upper part of the phantom are significantly increasingly suppressed with an increasing angular interval d. However, the image quality obtained with a full scan cannot be achieved, at least not with the present selection of d. Note that, in the noisy reconstruction, most of the artifacts are superimposed by the noise and are, therefore, not so clearly visible. Only in the short scan geometry with $d = 0°$ are some streaks close to the right ear recognizable.

Reconstruction error.   Figure 7.23 compares the weighting matrices in terms of reconstruction error measured within: i) the full background of the FORBILD head phantom (first row); and, ii) the specifically design elliptical region (second row), respectively. Table 7.4 lists the results for the bias measures obtained at the target resolution of $\nu_{0.5} = 4.75$/cm for the geometrical setting G3 in both pixel sizes.

For both MPS and SPS, the frequency-averaged bias measured in the full background of the phantom, $\langle \bar{\sigma}_b \rangle_\nu$, was clearly dependent on the geometry where the strongest effect was observed for geometries with FFS (G4 and G8). The frequency-averaged bias measured in the elliptical region close to the right ear, $\langle \bar{\sigma}_{ES} \rangle_\nu$, was also dependent on the geometrical setting but with less extent. This means the bias in

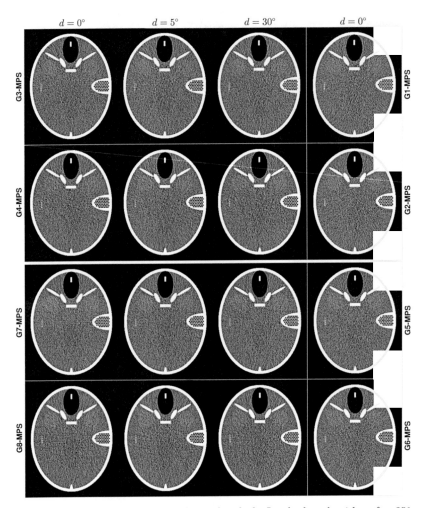

Figure 7.22: Noisy reconstructions obtained with the Landweber algorithm after 251 iterations for the short scan geometries G3-MPS, G4-MPS, G7-MPS, and G8-MPS *(first to fourth row)* with $\mathbf{W} = \mathbf{I}$ $(d = 0°)$, $\mathbf{W}_1$ $(d = 5°)$, and $\mathbf{W}_2$ $(d = 30°)$ *(first to third column)*. Noisy reconstructions of the corresponding full-scan geometry using $\mathbf{W} = \mathbf{I}$ $(d = 0°)$ *(fourth column)*; G1-MPS, G2-MPS, G5-MPS, and G6-MPS *(first to fourth row)*. Display window: c/w=50/200 HU.

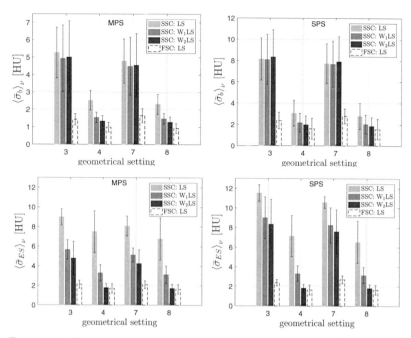

Figure 7.23: Frequency-averaged reconstruction error measures for MPS *(left)* and SPS *(right)*.

this region is mostly dominated by the definition of the start and stop angle of the short scan.

In most cases, a noticeable decrease in bias is achieved by including a data handling weighting matrix with $d \neq 0°$. For example, for MPS the bias $\langle \bar{\sigma}_{ES} \rangle_\nu$ was between 6.8 HU and 9.0 HU for $\mathbf{W} = \mathbf{I}$, between 3.1 HU and 5.7 HU for $\mathbf{W}_1$, and between 1.7 HU and 4.8 HU for $\mathbf{W}_2$; and for SPS, the bias was between 6.5 HU and 11.6 HU for $\mathbf{W} = \mathbf{I}$, between 3.2 HU and 9.0 HU for $\mathbf{W}_1$, and between 1.8 HU and 8.3 HU for $\mathbf{W}_2$. Note that with an increasing value of d bias is increasingly reduced, i.e., the maximum decrease in $\langle \bar{\sigma}_{ES} \rangle_\nu$ within one geometrical setting was up to 76.1% for MPS and up to 74.5% for SPS. The only exceptions were found for $\langle \bar{\sigma}_b \rangle_\nu$ in geometry G3 and G7 in MPS and SPS where the bias seem to pass through a minimum value of d that is between $d = 0°$ and $d = 30°$. That difference is not visible, however, in the noiseless reconstructions since the absolute value is miniscule, i.e., less than 0.08 HU for MPS and less than < 0.26 HU for SPS.

Apart from a significant reduction of the artifacts in the image, a data weighting matrix helps to speed up convergence in terms of necessary iterations, called N_{iter}, that are needed to reach a certain resolution (see Tab. 7.4). For example, the resolution of $\nu_{0.5} = 4.75/\text{cm}$ was reached 35 iterations earlier in MPS by using $d = 30°$

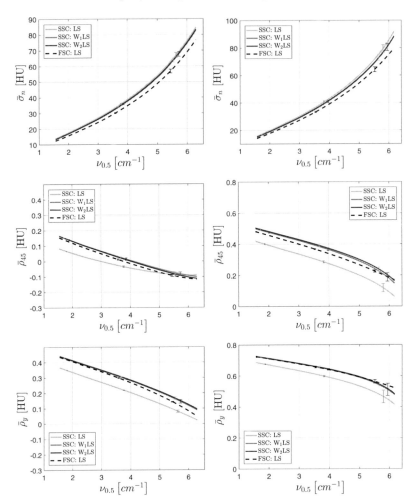

Figure 7.24: Geometry-averaged noise metrics for MPS *(left)* and SPS *(right)*. Mean standard deviation *(first row)*, correlation coefficient at 45 degrees *(second row)* and correlation coefficient in y *(third row)*.

instead of $d = 0°$. Nevertheless, the performance of the full scan was not reached in any of the geometrical settings.

Noise. Figure 7.24 compare the weighting matrices in terms of noise performance for MPS and SPS, respectively. The results for the correlation coefficient in x are not shown because they did not convey any additional information.

	N_{iter}	$\bar{\sigma}_b$ [HU]	$\bar{\sigma}_{ES}$ [HU]	$\bar{\sigma}_n$ [HU]	$\bar{\rho}_{45}$ [HU]	$\bar{\rho}_y$ [HU]
G3-MPS: SSC, $d = 0°$	316	5.54	7.45	48.89	−0.07	0.16
G3-MPS: SSC, $d = 5°$	286	5.57	5.80	47.47	−0.04	0.24
G3-MPS: SSC, $d = 30°$	281	5.75	5.32	47.78	−0.04	0.24
G3-MPS: FSC, $d = 0°$	256	1.32	1.83	44.40	−0.06	0.21
G3-SPS: SSC, $d = 0°$	296	8.93	13.18	53.74	0.22	0.56
G3-SPS: SSC, $d = 5°$	271	9.23	10.17	51.63	0.32	0.62
G3-SPS: SSC, $d = 30°$	266	9.61	9.83	51.83	0.33	0.62
G3-SPS: FSC, $d = 0°$	231	2.04	2.21	49.32	0.29	0.61

Table 7.4: Quantitative metric results for G3-MPS and G3-SPS at $\nu_{0.5} = 4.75/\text{cm}$. The parameter N_{iter} denotes the number of iterations needed to reach the target resolution.

For both, MPS and SPS, the dependence of noise metrics on frequency and geometry was very similar for all methods. Including a data redundancy handling matrix with $d \neq 0$ helps to reduce the mean standard deviation by up to 1.8 HU for MPS and by 3.7 HU for SPS, respectively. Note that $\mathbf{W}_1$ performed slightly better than $\mathbf{W}_2$. In comparison, the mean standard deviation of the full scan result was always at least 6 HU lower. The differences observed in the correlation coefficient were more significant. Using no weighting matrix ($\mathbf{W} = \mathbf{I}$) always yields the lowest correlation, while a data handling matrix with $d \neq 0$ shifted the correlation coefficients by up to 0.08 to higher values, bringing the correlation coefficients closer to the ones obtained for the full scan.

7.5.3.2   Influence of Statistical Weights and Penalty Term

Figure-of-merit.   All figures-of-merit represent the results obtained at the target resolution of $\nu_{0.5} = 4.75/\text{cm}$.

Display concept.   All results are displayed using a bar plot that shows the geometry specific result. Results corresponding to different metrics or pixel sizes are shown in separate figures, with the abbreviation MPS or SPS to emphasize to which pixel size a figure or plot corresponds. Each plot contains the result of the short scan (SSC) and the full scan (FSC) geometry for the three different groups of objective functions referred to as "$(\mathbf{W}_i)$LS", "$P(\mathbf{W}_i)$LS-QP" and "$P(\mathbf{W}_i)$LS-FP", where $(\mathbf{W}_i)$ indicates the absence or presence of a specific weighting matrix, called $\mathbf{W}_i$ with $i = 1, 2$. Three different tones of both contour and filling represent the different groups of objective functions: i) colored bars (contour and filling) are employed for the (weighted) least squares results; ii) gray colored bars (contour and filling) are employed for the penalized (weighted) least squares results using the quadratic potential; and, iii) transparent bars with gray colored contours are employed for the

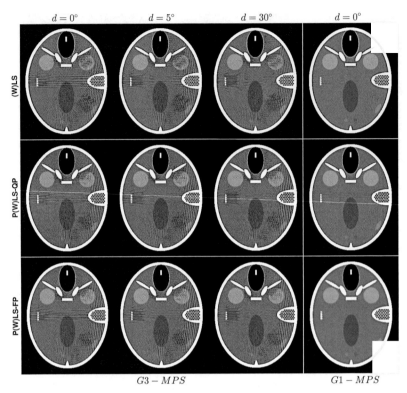

Figure 7.25: Noiseless reconstructions of the geometrical setting G3-MPS obtained from solving a (weighted) least squares problem *(first row)*, a penalized (weighted) least squares problem with quadratic penalty *(second row)*, and a penalized (weighted) least squares problem with the Fair potential *(third row)* under application from $\mathbf{W} = \mathbf{I}$ ($d = 0°$), $\mathbf{W}_1$ ($d = 5°$), and $\mathbf{W}_2$ ($d = 30°$) *(first to third column)*. Noiseless reconstructions of the corresponding full scan geometry G1-MPS using $\mathbf{W} = \mathbf{I}$ ($d = 0°$) *(fourth column)*. Display window: c/w=50/40 HU.

penalized (weighted) least squares results using the Fair potential. The full scan result is always displayed with a dashed contour line.

Visual appearance of some reconstructions. Figure 7.25 displays from top to bottom noiseless reconstructions of the geometrical setting G3-MPS obtained from solving a (weighted) least squares problem, a penalized (weighted) least squares problem with quadratic penalty, and a penalized (weighted) least squares problem with the Fair potential. The reconstructions in the first three columns correspond to a weighting matrix defined by $d = 0°, 5°, 30°$. Column four shows the corresponding full scan reconstructions of G1-MPS obtained with a weighting matrix defined by

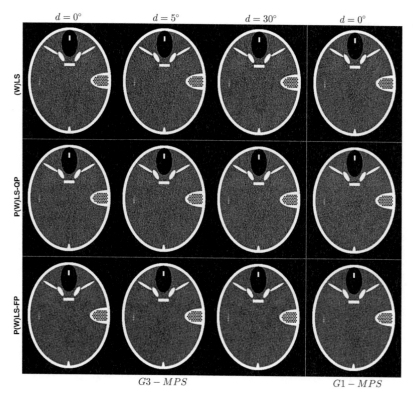

$$G3 - MPS \qquad\qquad\qquad G1 - MPS$$

Figure 7.26: Noisy reconstructions of the geometrical setting G3-MPS obtained from solving a (weighted) least squares problem *(first row)*, a penalized (weighted) least squares problem with quadratic penalty *(second row)*, and a penalized (weighted) least squares problem with the Fair potential *(third row)* under application from $\mathbf{W} = \mathbf{I}$ $(d = 0°)$, $\mathbf{W}_1$ $(d = 5°)$, and $\mathbf{W}_2$ $(d = 30°)$ *(first to third column)*. Noisy reconstructions of the corresponding full scan geometry G1-MPS using $\mathbf{W} = \mathbf{I}$ $(d = 0°)$ *(fourth column)*. Display window: c/w=50/200 HU.

$d = 0°$. Figure 7.26 displays noisy reconstructions for MPS in the same arrangement as in figure 7.25. Note that the grayscale window is more compressed for the noiseless reconstructions to emphasize the differences in image quality between the reconstructions. We only provide the reconstructions for one geometrical setting since the other reconstructions do not convey more information.

As clearly recognizable in the noiseless reconstructions, the image quality is significantly improved by introducing a data redundancy weight as part of the weighting matrix. The artifacts that appear close to the right ear or close to the slanted bones in the upper part of the phantom are particularly increasingly suppressed with an increasing angular interval d. In the noisy reconstruction, most of the artifacts are

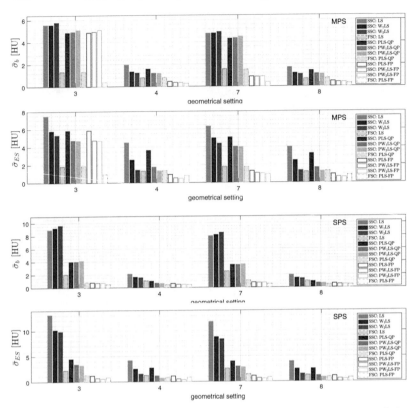

Figure 7.27: Reconstruction error measures $\bar{\sigma}_b$ and $\bar{\sigma}_{ES}$ for the three groups of objective functions including different Parker-like ray weighting matrices for MPS *(first and second row)* and SPS *(third and fourth row)*.

superimposed by the noise and are therefore not so clearly visible. For example, some streaks close to the right ear are only recognizable in the short scan reconstruction with $d = 0°$.

Reconstruction error. Figure 7.27 compares the weighting matrices and objective functions in terms of reconstruction error measured within: i) the full background of the FORBILD head phantom; and, ii) the specifically design elliptical region, referred to as $\bar{\sigma}_b$ and $\bar{\sigma}_{ES}$, respectively.

The objective function has a global impact on the nominal values obtained for the reconstruction error for both pixel sizes, whereas the trends across the three groups of objective functions remain largely the same. For example, the bias is reduced when a penalty term is used, with the biggest effect observed with the edge-preserving potential. Furthermore, in most cases a noticeable decrease in bias is achieved once a data redundancy weight with $d \neq 0°$ was included. The bias $\bar{\sigma}_{ES}$ over all (W)LS results was between 7.5 HU and 1.4 HU for MPS, and between 13.2 HU and 1.5 HU for SPS. In comparison, these values dropped over all P(W)LS results and were in the range between 5.8 HU and 0.5 HU for MPS, and between 4.5 HU and 0.4 HU for SPS.

Noise. Figure 7.28 compares the weighting matrices and objective functions in terms of noise performance for MPS and SPS, respectively. The results for the correlation coefficient in x are not shown because they did not convey any additional information.

As already indicated in the previous section, the objective function has a global impact on the nominal values obtained for each noise metric. The trend of the nominal values across the objective functions for the applied weighting matrices remains largely the same for all groups. However, the largest changes in the nominal values of the noise metrics were always observed for the edge-preserving potential.

In detail, a data redundancy weighting matrix with $d \neq 0°$ helps to reduce the mean standard deviation $\bar{\sigma}_n$ a little, specifically by up to 1.8 HU for MPS and by up to 2.2 HU for SPS, respectively. Thus, the weighting matrix brings the mean standard deviation closer to the one obtained for the full scan. Perhaps less predictable is the change on correlations. On the one hand, the results obtained with no weighting matrix change the correlation in 45-degree direction for MPS from a negative value to a positive value, i.e., for G3-MPS from -0.07 HU to 0.21 HU. For SPS, the pixels were already positively correlated in 45-degree direction. That correlation was enforced by up to 0.49 HU under a penalty term. The same effect is visible for the correlation in y-direction, where the positive correlation for MPS and SPS was again strengthened by the penalty term. On the other hand, the penalty term together with the data redundancy weight cause an even stronger correlation of the pixels, bringing the correlations closer to the one obtained for the full scan.

7.5.4   Summary Discussion

In the current part of the study we have presented the influence of data redundancy weights on image quality. The investigation was divided into two subsections. In the first subsection, the focus was on the influence of a statistical weighting matrix on its own, which was investigated using images created through Landweber iterations, which directly enabled observing the effects caused by the data redundancy weight. In the second subsection, a penalty term was added in the objective function. In total, two different data redundancy weights were applied: one with a small $(d = 5°)$ and one with a bigger $(d = 30°)$ angular interval over which the function $c(\lambda)$ drops smoothly from its maximum value of one to zero. For reasons of comparison, both a corresponding non-weighted $(d = 0°)$ short scan and full scan reconstructions for each geometrical setting were evaluated. The subsections revealed fundamental differences in terms of image quality. The salient observations were as follows:

- Even though a statistical iterative reconstruction method does not need a careful data redundancy handling concept for redundant data, the current study

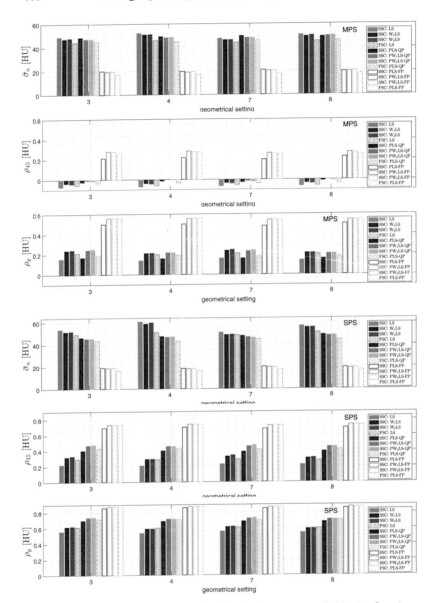

Figure 7.28: Noise measures $\bar{\sigma}_n$, $\bar{\rho}_{45}$ and $\bar{\rho}_y$ for the three groups of objective functions including different Parker-like ray weighting matrices for MPS *(first and third row)* and SPS *(fourth and sixth row)*.

showed that reconstruction errors may easily be reduced by having a weighting matrix modeling the data redundancies. The streak artifacts close to the right ear and close to the slanted small bones in the upper part of the phantom in particular were significantly reduced by up to 75%. As expected, a larger value of d removed the streaks more effective since the change in the weighting matrix is smoother than for a small d value. Furthermore, the reconstruction process was sped up by some tens of iterations for the WLS solution to reach the same target resolution.

- A penalty term further helps to improve the image quality compared to the PLS solution. Most of the remaining streaks were removed by the use of a data redundancy weight. The lowest bias and noise metrics were obtained for a penalized weighted least squares solution with the Fair potential. With the penalty term the bias close to the right ear could be additionally reduced by up to 25% and the noise was also reduced by up to 14%. In terms of correlation coefficients, both the penalty and the weighting matrix increased the binding between the pixels.

In summary, the use of a data redundancy weighting matrix helps to improve image quality. In this work already, small d-values suppressed the most dominant artifacts due to redundant data whereas the impact of such a weighting matrix on noise and correlations was very small. A regularization based on stopping after a given number of iterations, i.e., after 250 Landweber iterations, may be deemed satisfactory. However, the user should keep in mind that the design of the weighting matrix influences the final image resolution. In the given part of the study, the target resolution was reached faster by considering the data redundancy. Hence, in situations where the same resolution is desired, the reconstructed image needs to be post-smoothed. Under such circumstances, the penalized maximum-likelihood solution might be perceived as a more attractive reconstruction procedure.

7.6   Discussion and Conclusion

We presented an investigation of two different regularization approaches that are well known in iterative CT reconstruction. In the first approach, the regularization was based on stopping after a moderate number of iterations. That approach is popular in PET imaging since it generates an acceptable trade-off between resolution and noise. In the second approach, the regularization was directly enforced by a penalty term in the objective function. The solution of the first regularization approach was found through the Landweber algorithm, whereas the solution of the second approach was found by the application of the ICD method. Both regularization approaches were analyzed for two separate weighting matrices, whereas each of them covered an essential aspect in CT imaging. The first weighting matrix compromises non-uniform statistical weights by taking the mathematical model of a bowtie filter into account. The second weighting matrix reflects a data redundancy weight as it is often used in analytical CT reconstruction of short scan data. The influence of each weighting matrix related with the two regularization approaches were evaluated in separate parts. The focus of the first part was on the influence of the non-uniform statistical weights, whereas the focus of the second part was on the data redundancy weight. The study included a large variety of scanning geometries and a fair number of repeated scans in each geometry to assess dependence on photon statistics. The geometries were representative of contemporary CT scans. Together with the usage of two different image pixel sizes and a challenging phantom, a thorough investigation of the regularization approaches was possible.

The evaluation involved basic metrics covering resolution, discretization errors, image pixel noise, and statistical correlation between pixels. The metrics revealed fundamental differences between the two study parts. The first part of the study showed that a regularization based on the number of iterations might be very susceptible to reconstruction errors introduced by the design of the weighting matrix. Weighting non-smooth matrices particularly caused significant errors of several HU in the reconstructed images. Once a non-smooth weighting matrix was applied, iteration far beyond the desired resolution to remove the artifact was required. For the same set of weighting matrices, the reconstructions obtained by the ICD method were more stable. On the other hand, the impact on the noise-resolution trade-off was less predictable for both regularization techniques. However, the resolution of the images was reached more quickly, i.e., the necessary iterations to reach a given resolution was reached by some tens of iterations faster using the Landweber algorithm. Similar results were obtained in the second part of the study. The data redundancy weighting matrix was a smooth function that clearly helped to reduce reconstruction errors caused by data redundancy. Although the ICD method was again more stable in the reconstruction process in terms of reconstruction errors, the inclusion of the weighting matrix still helped to reduce bias. In terms of the noise metrics, it was possible to bring the results closer to the ones obtained for the full scan. As in the first part, inclusion of a weighting matrix again helped to speed up the reconstruction process when using the Landweber algorithm.

From our results, we can make two salient observations: First, it is not straightforward to state that a weighting matrix always helps to reduce reconstruction errors even though it models the physics better. Sometimes, such a weighting matrix might introduce undesirable side-effects in the image reconstruction, such as additional, nonresolution-related reconstruction errors or unpredictable noise effects. If a weighting matrix is desired, the weighting function should be as smooth as possible, which in turn might be a very challenging task. However, before including such a weighting matrix, it seems to be very important to understand how such a weighting matrix influences the final image quality. Hence, a detailed investigation of the influence of the weights might help to even speed up and improve the final image reconstruction of any iterative reconstruction algorithm. Second, the study shows that the noise-resolution trade-off introduced by the iteration number is not always attractive for CT imaging, particularly when statistical weights and data redundancies are involved. In such situations, the penalized (weighted) maximum-likelihood solution might be perceived as a more attractive procedure. An alternative approach might be to initialize the reconstruction process with a filtered-backprojection procedure. However, in this case, it is important to understand what component of this first image remains when the reconstruction is completed.

CHAPTER 8

FINESSE: a Fast Iterative Non-linear Exact Sub-space SEarch based Algorithm

This chapter presents a novel iterative reconstruction algorithm for CT imaging, called FINESSE and is structured as follows: Section 8.1 discusses the motivation of the development of a new iterative reconstruction algorithm. Next, in Section 8.2, some background and the algorithm design of FINESSE is described. This section is followed by Section 8.3, which contains information about the simulation study, e.g., information about the phantom selection, data simulation, forward projection model, the quantitative evaluation details, the results, and a summary. First real data results are shown in Section 8.4. The chapter concludes with a discussion and conclusion in Section 8.5.

Parts of this work have already been published in Schmitt et al. [Schm 14b]. The algorithm has also been patented by Schmitt et al. [Schm 17].

8.1 Motivation and Clinical Application

Statistical iterative reconstruction methods have become a subject of intense research primarily because they may allow significantly reducing radiation dose to the patient [Pick 12, Desa 12, Nero 13, Vard 12, Hara 09, Fahi 13]. Several reconstruction algorithms based on the penalized weighted least squares model have been suggested [Fess 94, Wang 06].

Most existing algorithms can be classified according to the number of voxels that are updated within one iteration. For example, the ICD method [Abat 82, Luo 92, Thib 07, Yu 11] updates one voxel at a time, yielding an algorithm that converges quickly, but is poorly parallelizable and, thus, time consuming. Other methods, such as the ordered subsets (OS) algorithms [Huds 94] based on separable quadratic surrogates (SQS) [Kamp 98, Erdo 99] or (preconditioned) nonlinear conjugate gradient (NCG) methods [Fess 99, Noce 99], update all voxels simultaneously and are therefore much more amenable to parallelization; however these algorithms require many more iterations than ICD. Another attractive algorithm that updates all voxels simultaneously is the fast iterative shrinkage-thresholding algorithm (FISTA) [Daub 04, Beck 09]. That algorithm has a fast outer converging loop, but is hampered by the need to solve a non-linear image-denoising problem with correlated voxels at each iteration. Last, there are block-based coordinate descent methods that are meant to update several, but not all, voxels within each iteration [Bens 10, Fess 97, Fess 11]. Like FISTA, these algorithms suffer from the drawback of requiring the solution of a complex sub-problem, which is often achieved using an approximate, monotonic update that slows down convergence.

Currently, none of these algorithms is seen as being satisfactory for routine clinical usage. An ideal and practical iterative reconstruction algorithm would satisfy the desired properties of fast and global convergence, of being stable and convergent, and of being fast, robust, parallelizable, and flexible. With the development of FINESSE, we try to realize most of these properties. This is achieved by integrating together the principles of FISTA and block-based coordinate descent methods to retain the advantages of each method while overcoming their deficiencies.

8.2 Algorithm Design

In this section the algorithm design of FINESSE is presented. Some background information about the problem formulation and the description of FISTA is given before explaining the general algorithm concept.

8.2.1 Problem Formulation

FINESSE is based on the penalized weighted least squares model, i.e., the algorithm seeks the solution of a constrained image reconstruction problem as defined in Section 4.2.2:

$$\boldsymbol{f}^* = \arg\min_{\boldsymbol{f}} \Phi(\boldsymbol{f}, \boldsymbol{g}) \ , \qquad \text{where}$$

$$\Phi(\boldsymbol{f}, \boldsymbol{g}) = \Phi_M(\boldsymbol{f}, \boldsymbol{g}) + \Phi_R(\boldsymbol{f}) \ .$$

The investigation of the algorithm is restricted to expressions of $\Phi_R(\boldsymbol{f})$ that allow $\Phi(\boldsymbol{f}, \boldsymbol{g})$ to be convex. Moreover, in our study $\Phi_R(\boldsymbol{f})$ only links the voxels to their immediate neighbors.

8.2.2 Description of FISTA

FISTA is a fast converging algorithm that has been extensively studied in the literature. Let L be the Lipschitz constant for $\Phi_M(\boldsymbol{f}, \tilde{\boldsymbol{g}})$, where $\Phi_M(\boldsymbol{f}, \tilde{\boldsymbol{g}}) = \left\| \tilde{\mathbf{A}} \boldsymbol{f} - \tilde{\boldsymbol{g}} \right\|_2^2$, let $\tilde{\boldsymbol{f}}$ be the result of a very specific linear combination of the previous two iteration

---

Algorithm 8.1: FISTA with constant step size

Input: $\mathbf{W}$, $\mathbf{g}$, $\tilde{\boldsymbol{f}}^{(1)} = \boldsymbol{f}^{(0)}$ with $\boldsymbol{f}^{(0)} \neq 0$, $t^{(1)} = 1$, choice of the forward projection model [a] and the regularization function [b]

Output: reconstructed image volume $\boldsymbol{f}$

// pre-computations before starting with the first iteration
1   compute projector elements a_{rs}
2   apply the statistical weights, i.e., compute: $\tilde{\mathbf{A}} = \mathbf{W}\mathbf{A}$ and $\tilde{\boldsymbol{g}} = \mathbf{W}\boldsymbol{g}$
3   compute: $\tilde{\mathbf{A}}\boldsymbol{f}^{(0)}$
4   compute: Lipschitz constant L

// FISTA iterations
5   **for** $n = 1, 2, \ldots$ **do**
6   calculate $\boldsymbol{z}^{(n)}$ (Eqn. 8.5)
7   calculate $\boldsymbol{f}^{(n)}$ by solving Eqn. 8.2
8   update parameter t according to Eqn. 8.3
9   update $\tilde{\boldsymbol{f}}$ according to Eqn. 8.4
10   *repeat the previous steps until convergence is achieved* [c]
11 **end**

---

[a] The following forward projection models have been implemented: Joseph's method, distance-driven method, and B-splines (see Sec. 6.2);
[b] The following potential functions and their corresponding derivations have been implemented: Fair potential and a quadratic potential.
[c] The exact stopping criterion is defined by the user.

---

results $\left\{ \tilde{\boldsymbol{f}}^{(n-1)}, \tilde{\boldsymbol{f}}^{(n-2)} \right\}$ and let $t^{(n)}$ be a parameter that depends on the iteration n. Then, FISTA with constant step size generates the following sequence of iterates:

$$z^{(n)} = \tilde{\boldsymbol{f}}^{(n)} - \frac{1}{L}\nabla\Phi_M(\tilde{\boldsymbol{f}}^{(n)}, \tilde{\boldsymbol{g}}) \; , \tag{8.1}$$

$$\boldsymbol{f}^{(n)} = \arg\min_{\boldsymbol{f}} \left\{ \frac{L}{2} \left\| \boldsymbol{f} - \boldsymbol{z}^{(n)} \right\|^2 + \Phi_R(\boldsymbol{f}) \right\} \; , \tag{8.2}$$

$$t^{(n+1)} = \frac{1 + \sqrt{1 + 4 \cdot \left(t^{(n)}\right)^2}}{2} \; , \tag{8.3}$$

$$\tilde{\boldsymbol{f}}^{(n+1)} = \boldsymbol{f}^{(n)} + \left(\frac{t^{(n)} - 1}{t^{(n+1)}} \right) \left(\boldsymbol{f}^{(n)} - \boldsymbol{f}^{(n-1)} \right) \; . \tag{8.4}$$

When $\Phi_M(\tilde{\boldsymbol{f}}^{(n)}, \tilde{\boldsymbol{g}}) = \left\| \tilde{\mathbf{A}}\tilde{\boldsymbol{f}}^{(n)} - \tilde{\boldsymbol{g}} \right\|_2^2$ Equation 8.1 may also be expressed as

$$z^{(n)} = \tilde{\boldsymbol{f}}^{(n)} - \frac{2}{L}\tilde{\mathbf{A}}^T \left(\tilde{\mathbf{A}}\tilde{\boldsymbol{f}}^{(n)} - \tilde{\boldsymbol{g}} \right) \; , \tag{8.5}$$

which is a Landweber style update step. The algorithm is usually initialized with $t^{(1)} = 1$ and $\tilde{\boldsymbol{f}}^{(1)} = \tilde{\boldsymbol{f}}^{(0)}$. A detailed overview of the single steps of FISTA is shown in Algorithm 8.1.

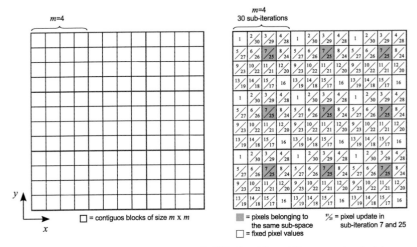

Figure 8.1: General concept of the FINESSE algorithm. The reconstructed image volume x is divided into contiguous blocks of equal edge length m *(left)* . The figures in each block define the ordering of the sub-spaces *(right)*. Voxels belonging to the same sub-space, i.e., green highlighted pixels, are updated at the same time while all the other voxel values are being fixed during the update step.

The outer converging loop of FISTA is fast, however, the minimizing step $f^{(n)}$ is very time consuming and complicated to solve, since all pixels are updated at the same time. Moreover, the full projection matrix $\tilde{A}$ needs to be fully available which may require a lot of memory.

8.2.3 Description of FINESSE

8.2.3.1 General Concept

The FINESSE algorithm updates the voxels in groups. The reconstructed image volume x is divided into contiguous blocks of equal edge length m. Figure 8.1 displays this concept on the left. The group of voxels that is simultaneously updated is defined by selecting one element out of each block, using the same element position from one block to the next. The voxels within one group define a *subspace*, e.g., green highlighted pixels on the right hand side in Figure 8.1. These voxels are updated by exactly solving the optimization problem corresponding to minimizing $\Phi(f, g)$, while keeping the other voxels fixed. This sub-optimization problem is solved using the fast iterative steps of FISTA. Given that $\Phi_R(f)$ only links one voxel to its immediate neighbors and these neighbors are fixed, the image denoising problem given by Equation 8.2 in FISTA reduces to a stack of independent, one-dimensional optimization problems that can be solved in parallel using, for instance, the Newton-Raphson method, that is quadratically converging.

The group of voxels that are updated simultaneously define a *subspace*. An update step of all voxels within one group is called *sub-iteration*. A *full iteration* has been performed when all defined sub-iterations are completed. As well known from literature [Kim 15], the ordering of the sub-iterations affects convergence. A popular ordering of the sub-spaces is the random or the back-and-forth approach. In the random approach, the voxels are updated in a random order. In the back-and-forth scheme, the first sub-iteration belongs to a corner voxel of the block. All voxel values are updated one after another until the last voxel of the block is reached, then all voxels are again updated by going back towards the first voxel in the block, in reverse order.

8.2.3.2   Mathematical Description

With the properties of FINESSE as described above, the following mathematical description applies. Let s define the s-th subspace of altogether N_s subspaces, let $\Box_s$ be a placeholder for a subspace matrix or a subspace function, and let $\bar{\Box}_s$ be the complement of the s-th subspace matrix or subspace function. Then, the reconstruction result $\boldsymbol{f}$ and the projection matrix $\tilde{\mathbf{A}}$ can be expressed as

$$\boldsymbol{f} = \left[\boldsymbol{f}_s, \, \bar{\boldsymbol{f}}_s \right]^T \quad \text{and} \quad \tilde{\mathbf{A}} = \left[\tilde{\mathbf{A}}_s, \, \bar{\tilde{\mathbf{A}}}_s \right] . \tag{8.6}$$

Using the definitions given above the objective function $\Phi(\boldsymbol{f}, \tilde{g})$ can be rewritten as follows

$$\Phi(\boldsymbol{f}, \tilde{g}) = \left\| \tilde{\mathbf{A}} \boldsymbol{f} - \tilde{g} \right\|_2^2 + \Phi_R(\boldsymbol{f}) \tag{8.7}$$

$$= \left\| \tilde{\mathbf{A}}_s \boldsymbol{f}_s + \bar{\tilde{\mathbf{A}}}_s \bar{\boldsymbol{f}}_s - \tilde{g} \right\|_2^2 + \Phi_R(\boldsymbol{f}_s, \bar{\boldsymbol{f}}_s) . \tag{8.8}$$

Now, let $\tilde{g}_s = \tilde{g} - \bar{\tilde{\mathbf{A}}}_s \bar{\boldsymbol{f}}_s$. Further, let l_s, with $l_s = 1, 2, \ldots, N_{l_s}$, count the voxels of the subspace s and let k, with $k = 1, \ldots, N_k$, define the direct neighbor pixels of the voxel f_{s, l_s}. Then, the regularization term of the subspace s, called $\Phi_{R,s}$, is given by

$$\Phi_{R,s}(\boldsymbol{f}_s, \bar{\boldsymbol{f}}_s) = \beta_R \sum_{l_s}^{N_{l_s}} \sum_{k}^{N_k} \omega_{l_s k}^{-1} \, \phi(f_{s, l_s} - \bar{f}_{s,k}) . \tag{8.9}$$

Finally, the subspace specific formulation of the objective function is

$$\Phi(\boldsymbol{f}_s, \tilde{g}_s) = \left\| \tilde{\mathbf{A}}_s \boldsymbol{f}_s - \tilde{g}_s \right\|_2^2 + \beta_R \sum_{l_s}^{N_{l_s}} \sum_{k}^{N_k} \omega_{l_s k}^{-1} \, \phi(f_{s, l_s} - \bar{f}_{s,k}) . \tag{8.10}$$

Note that ϕ is a local (nearest neighbor) function and since the voxel values $\bar{f}_{s,k}$ are unchanged during the update step of the subspace s, the term $\Phi(\boldsymbol{f}_s, \tilde{g}_s)$ becomes a separable quantity. This means that the regularization term may be expressed as a sum: $\Phi_{R,s}(\boldsymbol{f}_s) = \sum_{l_s} \Phi_{R,s}(f_{s, l_s})$.

Each subspace optimization problem is now solved using the general steps of FISTA:

$$z_s^{(n_s)} = \tilde{f}_s^{(n_s)} - \frac{2}{L_s}\tilde{A}_s^T\left(\tilde{A}_s\tilde{f}_s^{(n_s)} - \tilde{g}_s\right), \tag{8.11}$$

$$f_s^{(n_s)} = \arg\min_{f_s}\left\{\frac{L_s}{2}\left\|f_s - z_s^{(n_s)}\right\|^2 + \Phi_{R,s}(f_s)\right\}, \tag{8.12}$$

$$t_s^{(n_s+1)} = \frac{1 + \sqrt{1 + 4\cdot\left(t_s^{(n_s)}\right)^2}}{2}, \tag{8.13}$$

$$\tilde{f}_s^{(n_s+1)} = f_s^{(n_s)} + \left(\frac{t_s^{(n_s)} - 1}{t_s^{(n_s+1)}}\right)\left(f_s^{(n_s)} - f_s^{(n_s-1)}\right). \tag{8.14}$$

Note that in FINESSE two iteration counters exist: n and n_s. While n_s will be always reset to one for a new subspace optimization problem, the counter n continuously increases by one when all defined subspaces have been optimized. The algorithm is initialized with $\tilde{f}^{(1)} = \tilde{f}^{(0)}$. For each first subspace optimization step, the update parameter t_s needs to be reset to $t_s^{(1)} = 1$. Equation 8.11 is again a Landweber style update step. As mentioned earlier, Equation 8.12 reduces to a stack of independent, one-dimensional optimization problems that can be solved in parallel using, for instance, the Newton-Raphson method.

Due to the finite computational precision a threshold parameter, called T^{NRM}, is introduced to avoid a possible "ping-pong" optimization effect. That parameter defines the exactness of the Newton-Raphson method. When the change of the pixel value, $\Delta f_{s,l_s}^{NRM}$, is smaller than T^{NRM}, the Newton-Raphson optimization loop is terminated. Furthermore, a second threshold parameter, T^{OF}, defines the upper limit of the change in the objective function value, $\Delta\epsilon^{OF}$, where $\Delta\epsilon^{OF} = \Phi(f_s^{(n_s+1)}, \tilde{g}_s) - \Phi(f_s^{(n_s)}, \tilde{g}_s)$. When $\Delta\epsilon^{OF} < T^{OF}$, the sub-iteration is terminated and the optimization of the next subspace will be started.

A detailed overview of the single steps of FINESSE is given in Algorithm 8.2.

8.2.3.3   What Sets FINESSE Apart From Other Algorithms

The design of FINESSE has at least two compelling advantages. First, the optimization problem can be solved exactly. Second, the image denoising problem can be solved in parallel, which is much more attractive than that encountered in the classical FISTA algorithm exposed above.

Reduction of the optimization problem to a stack of one-dimensional problems that can be solved in parallel is a feature that is seen in other block-coordinate descent methods, but there is a fundamental difference between these methods and FINESSE: in FINESSE that feature is achieved while performing exact subspace search. Thereby, FINESSE integrates together the principles of FISTA and block-based coordinate descent methods to retain the advantages of each method while overcoming their deficiencies. For these reasons, we named our new algorithm FINESSE: fast iterative non-linear exact sub-space search based algorithm.

Algorithm 8.2: FINESSE algorithm

Input: $\mathbf{W}$, $\mathbf{g}$, choice of the forward projection model[*], regularization function[**]
$\tilde{\boldsymbol{f}}^{(1)} = \boldsymbol{f}^{(0)}$ with $\boldsymbol{f}^{(0)} \neq 0$, $t_s^{(1)} = 1$, block size m,
arrangement of the subspaces (order schema), thresholds[***] T^{OF}, T^{NRM}
Output: reconstructed image volume $\boldsymbol{f}$

// pre-computations before starting with the first iteration

1 compute projector elements a_{rs}
2 apply the statistical weights, i.e., compute: $\tilde{\mathbf{A}} = \mathbf{W}\mathbf{A}$ and $\tilde{g} = \mathbf{W}g$
3 compute: $\tilde{\mathbf{A}}\boldsymbol{f}^{(0)}$
4 compute Lipschitz constants: L_s
5 compute objective function value ϵ^{OF}

// FINESSE iterations

6 **for** $n = 1, 2, \ldots$ **do**   (full iterations)
7   **for** $s = 1, \ldots, N_s$ **do**   (solve subspace min. problem using FISTA)
8   set $t_s^{(1)} = 1$
9   **while** $\Delta\epsilon^{OF} > T^{OF}$
10   compute Landweber style update step: $\boldsymbol{z}_s^{(n_s)}$ (Eqn. 8.11)
11   **parallel for** $l_s = 1, \ldots, N_{l_s}$ (solve min. problem in parallel)
12   **while** $\Delta f_{s,l_s}^{NRM} > T^{NRM}$
13   exact subspace search: Newton-Raphson method
14   **end**
15   **end**
16   update parameter t_s (Eqn. 8.13)
17   update parameter $\tilde{\boldsymbol{f}}_s$ (Eqn. 8.14)
18   compute $\Delta\epsilon^{OF}$
19   **end**
20   **end**
21 **end**

---

[*] The following forward projection models have been implemented: Joseph's method, distance-driven method, and B-splines (see Sec. 6.2);
[**] The following potential functions and their corresponding derivations have been implemented: Fair potential and a quadratic potential.
[***] The exact stopping criterion is defined by the user.

8.3 Simulation Study

8.3.1 Experimental Setup

8.3.1.1 Geometrical Settings

A selection of 8 geometrical (parametric) settings as defined in Section 6.3.2 were used in the simulation study. The focus here was on the geometrical settings without FFS, i.e., on G1, G3, G5, and G7 in moderate and small pixel size. The label concept and the parameter settings for each geometry are given on page 67.

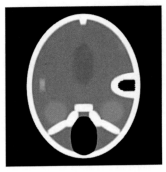

Figure 8.2: Illustration of a smooth FBP reconstruction used for initialization of FINESSE. Display window: c/w=50/200 HU.

8.3.1.2 Phantom Selection and Data Simulation

The simulations were carried out using the FORBILD head phantom. All data simulations were performed in fan-beam geometry for a 3rd generation CT scanner with a curved detector. In detail, the same concept of modeling the rays with the same modeling parameters as described in Section 6.3.3 were used in this study. This means that, due to a sub-sampling of the focal spot, each detector element, each view, and each ray was modeled as an analytical, non-linear average of 405 line integrals. In total, 50 noise realizations for each geometrical setting and each phantom were available. All data simulation and image reconstruction parameters are listed in Table 6.1.

8.3.1.3 FINESSE Settings

The forward projection matrix $\mathbf{A}$ was formed using the distance-driven method. That choice was related to the results obtained in Chapter 6. Moreover, to avoid any potential distortions in the evaluation of the reconstruction algorithm, the statistical weighting matrix was the identical matrix, i.e., $\mathbf{W} = \mathbf{I}$. For each sub-iteration, parameter L_s can be shown to be equal to $2\,\sigma_{max}$, where σ_{max} is the maximum singular value of the sub-projection matrix; σ_{max} and, thereby, L_s, was estimated using the Power method [Golu 96].

The algorithm was always initialized with a smooth FBP reconstruction. Figure 8.2 shows an example of that smooth FBP image. The step size parameter t was always initialized with one.

The regularization term was formed using the quadratic potential with the parameter ϵ being 500. As in the previous chapter, the target resolution of the phantom study was set to $\nu_{0.5} = 4.75/\text{cm}$, which is a moderate resolution for brain imaging. This means that, for the phantom study, due to various geometrical settings different regularization parameters β_R are essential to attain the given resolution (see Sec. 7.3). The regularization parameters used are given in Table 8.1.

geometrical setting			
G1	G3	G5	G7
MPS	0.4332	0.2313	0.4101	0.2170
SPS	0.3180	0.1639	0.2988	0.1514

Table 8.1: Regularization parameter value settings used in the simulation study.

A sub-iterations was stopped when the difference in the objective function became smaller than $T^{OF} = 10^{-5}$, with the embedded Newton-Raphson method being stopped when the change of the pixel value was less than $T^{NRM} = 10^{-6}$ HU.

Three different block sizes were investigated, namely $m = 2, 3, 4$. The sub-spaces were sort using the back-and-forth approach. Specifically, the first sub-iteration was placed on the upper left corner of the block, then one voxel value after another were updated until the last voxel in the block was reached, and then repeated back towards the first voxel in the block, in reverse order. Hence, a full iteration consists of $2m^2 - 2$ sub-iterations (see illustration in Fig. 8.1).

Note that due to the settings defined above the output of FINESSE results in exactly one reconstructed image.

8.3.1.4   Reconstruction Methods

FINESSE was compared against two other reconstruction algorithms: i) the ICD method using a quadratic regularization; and, ii) the Landweber algorithm. Both algorithms were applied without application of a statistical weighting matrix, i.e., $\mathbf{W} = \mathbf{I}$. For the Landweber result, the iterate that corresponds to the target resolution of $\nu_{0.5} = 4.75/cm$ was used. For the ICD method, the geometrical and pixel size dependent regularization parameters β_R as given in Table 8.1 were applied.

For a simplified assignment of the results, we introduce the following labeling concept which is based on the labeling concept of the previous chapter. Results obtained with the ICD method will be labeled with "PLS-ICD", whereas the results of the Landweber algorithm will be labeled with "LS-LW"[1]. Moreover, the results of FINESSE will be labeled with "PLS-F: $m = b$" with $b = 2, 3, 4$ defining the block size used in the reconstruction.

8.3.1.5   Image Quality Assessment

Image quality was assessed using basic metrics. In detail, the mean absolute reconstruction error (bias) and noise was evaluated as defined in Section 5.2.4 and 5.2.5, respectively. Since close to the right ear high reconstruction errors are expected, especially for the short scan geometries, the bias was additionally calculated in a specifically defined region as displayed on the right side in Figure 7.7. The definition of calculating the bias for the evaluation regions is defined in Equation 5.3. To

---

[1]Our labeling concept intentionally neglects the fact that the Landweber algorithm was quasi-penalized by the application of a fixed number of iterates.

avoid confusion and to be consistent with the previous chapter, the bias corresponding to the highlighted region in Figure 7.7 is called $\bar{\sigma}_{ES}$, whereas the bias of the full background of the phantom is called $\bar{\sigma}_b$.

8.3.1.6   Computational Cost

Iterative reconstruction methods often suffer from the computational effort. The measurement of computational cost in terms of time depends strongly on the implementation. To overcome this issue, which was not a topic of this dissertation, the metric computational cost, called N_{iter}, was limited to the total number of necessary iterations to reach the target resolution. The metric was only applied to the noiseless reconstruction.

8.3.2   Results

Figure-of-merit. All figures-of-merit represent the results obtained at the target resolution of $\nu_{0.5} = 4.75/\mathrm{cm}$.

Display concept. All basic metric results are displayed using a bar plot that shows the geometry specific result. Results corresponding to different metrics or pixel sizes are shown in separate figures, with the abbreviation MPS or SPS to emphasize to which pixel size a figure or plot corresponds. Each plot contains the result of the three different block sizes of FINESSE referred to as "PLS-F: $m = 2$", "PLS-F: $m = 3$", and "PLS-F: $m = 4$" followed by the result of the ICD method referred to as "PLS-ICD" and by the result of the Landweber algorithm referred to as "LS-LW". Two different tones of both contour and filling represent the different groups of applied reconstruction algorithms: i) colored bars (contour and filling) are employed for the FINESSE results; and, ii) transparent bars with dark and light gray colored contours are employed for the ICD method and the Landweber algorithm, respectively.

Visual appearance of some reconstructions. Figure 8.3 displays from top to bottom noiseless reconstructions of the geometrical settings G1-SPS, G3-SPS, G5-SPS and G7-SPS obtained from (left to right) FINESSE with a block size of $m = 2$ and $m = 3$, from the ICD method and the Landweber algorithm. Figure 8.4 displays noisy reconstructions for SPS in the same arrangement as in Figure 8.3. Note that the grayscale window is more compressed for the noiseless reconstructions to emphasize the differences in image quality between the reconstructions. We only provide the reconstructions for FINESSE in small pixel size with block sizes $m = 2$ and $m = 3$, setting since the moderate pixel size and the block size of $m = 4$ do not convey any more information. Figure 8.5 shows the difference image PLS-ICD – PLS-F: $m = 2$, PLS-ICD – PLS-F: $m = 3$ and PLS-ICD – LS-LW obtained from the noiseless reconstruction of G3-SPS.

The noiseless reconstructions show nicely the difference in image quality. FINESSE reconstructions are relatively comparable in terms of image quality with the ICD results but are not as "smooth" as reconstruction obtained with the ICD method;

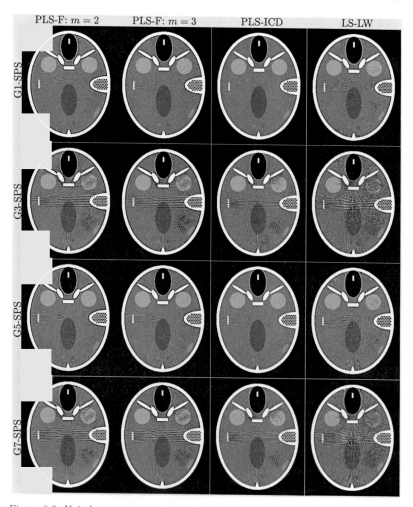

Figure 8.3: Noiseless reconstructions of the geometrical settings G1, G3, G5 and G7 in small pixel size *(top to bottom)*. The reconstructions were obtained with FINESSE using a block size of $m = 2$ and $m = 3$, with the ICD method using the quadratic potential (PLS-QP) and with the Landweber algorithm (LS) *(left to right)*. Display window: c/w=50/40 HU.

the short scan reconstructions particularly show much more prominent short scan artifacts (see also Fig. 8.5). On the other hand, the full scan geometries (G1 and G5) do not show such obvious differences in image quality. The Landweber algorithm results in reconstructions, which all suffer from high frequency and short scan artifacts. In

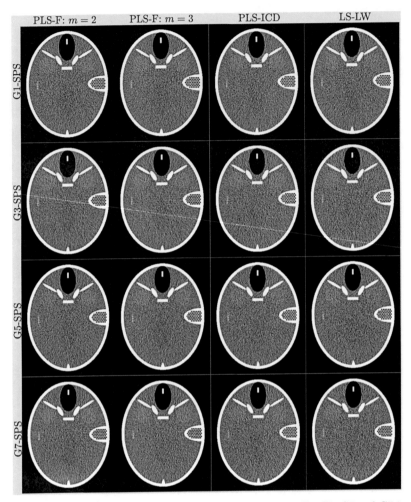

Figure 8.4: Noisy reconstructions of the geometrical settings G1, G3, G5 and G7 in small pixel size *(top to bottom)*. The reconstructions were obtained with FINESSE using a block size of $m = 2$ and $m = 3$, with the ICD method using the quadratic potential (PLS-QP) and with the Landweber algorithm (LS) *(left to right)*. Display window: c/w=50/200 HU.

the noisy reconstructions most of the artifacts are superimposed by the noise and are therefore not so clearly visible.

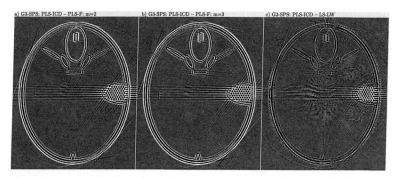

Figure 8.5: Difference images for the noiseless reconstruction for G3-SPS: *a)* PLS-ICD – PLS-F: m=2, *b)* PLS-ICD – PLS-F: m=3, and *c)* PLS-ICD – LS-LW. Display window: c/w=0/50 HU.

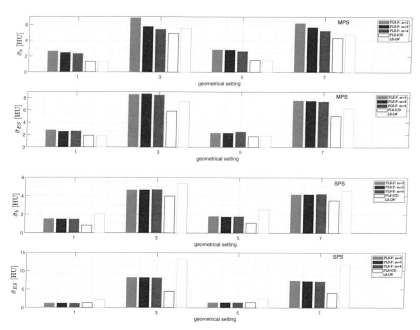

Figure 8.6: Reconstruction error measures $\bar{\sigma}_b$ and $\bar{\sigma}_{ES}$ for the three reconstruction algorithms for MPS *(first and second row)* and SPS *(third and fourth row)*.

Reconstruction error.   Figure 8.6 compares the reconstruction methods for both pixel sizes in terms of reconstruction error measured within: i) the full background of the FORBILD head phantom; and, ii) within the elliptical-shaped ROI within the FORBILD head phantom, referred to as $\bar{\sigma}_b$ and $\bar{\sigma}_{ES}$, respectively.

The reconstruction method has a global impact on the nominal values obtained for the reconstruction error for both pixel sizes. For MPS, the measured reconstruction error of FINESSE is always a little higher than the error measured for the ICD method or with the Landweber algorithm. For SPS the ICD method give the lowest reconstruction error results followed by FINESSE and Landweber. As the noiseless reconstruction already indicated, these differences in bias are mostly driven by the streak artifacts.

The FINESSE settings have also a miniscule influence on the result of the bias especially when a moderate pixel size is selected. In comparison to a block size of $m = 4$, the bias $\bar{\sigma}_b$ for $m = 3$ increases between 5.0% and 8.5% and for $m = 2$ between 7.7% up to 18.6%, whereas, for SPS, $\bar{\sigma}_b$ is more or less constant (less than 1% differences) over the block sizes.

Noise. Figure 8.7 compares the reconstruction methods for both pixel sizes in terms of noise measures. The results for the correlation coefficient in x are not shown because they did not convey any additional information.

As already indicated in the previous paragraph, the reconstruction algorithm has a global impact on the nominal values obtained for each noise metric. For MPS, the nominal value of the noise measure $\bar{\sigma}_n$ depends on the scan mode. For the full scan geometries, the noise with FINESSE was up to 30% higher than with ICD, whereas in the short scan geometries the noise was up to 17% less than with ICD. The differences in correlation between pixels also seem to be dependent on the scan method. For instance, in FINESSE the correlation ρ_{45} is a little stronger for the full scan geometries and less for short scan geometries. For the correlation ρ_y, that trend is inverse compared to the correlation ρ_{45}. For SPS, the noise metric $\bar{\sigma}_n$ for FINESSE is always lower than for the ICD or Landweber method and the correlation coefficients of FINESSE are in the same range as for the ICD method.

The FINESSE settings also have an influence on the result of the noise measures, especially when a moderate pixel size is selected (same as for the bias measures). For MPS, the noise measure $\bar{\sigma}_n$ decreases up to 17% when a block size of $m = 4$ instead of $m = 2$, is used. The correlation in xy-direction decreases up to 46% for using $m = 4$ instead of $m = 2$ whereas for that setting the correlation in y-direction increases up to 40%. For SPS, all noise measures are more or less constant, i.e., the variation is less than 1%.

Computational Cost. Figure 8.8 shows the computational cost metric for all reconstruction methods for both pixel sizes. FINESSE needs many fewer iterations than the ICD method and the Landweber algorithm. That comparison is limited to the fact that FINESSE has quite a lot of sub-spaces with a moderate number of sub-iterations. However, to count the number of sub-iterations within a sub-space of FINESSE will not be a fair comparison, since the ICD method also needs to perform a half-interval search that also costs iterates. To overcome that issue, the number N_{iter} was multiplied with the number of sub-spaces, i.e., $N_{iter} \cdot (2m^2 - 2)$, while N_{iter} was unchanged for the ICD method and the Landweber algorithm. Table 8.2 lists

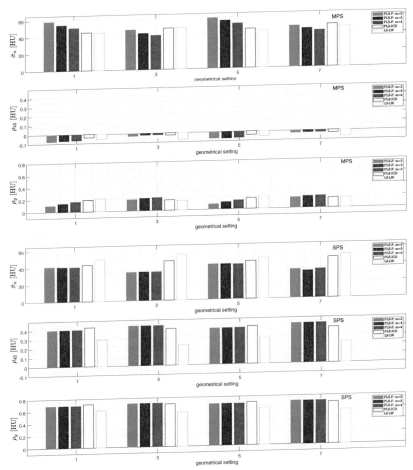

Figure 8.7: Noise measures $\bar{\sigma}_n$, $\bar{\rho}_{45}$ and $\bar{\rho}_y$ for the three reconstruction algorithms for MPS *(first to third row)* and SPS *(fourth to sixth row)*.

the modified number of iterations. That table shows that except one case (G7-MPS) the number of iterations of FINESSE is still always lower than for ICD.

8.3.3   Summary Discussion

In this section, we have presented the reconstruction results of our novel iterative reconstruction algorithm FINESSE together with a comparison against two other iterative reconstruction algorithms. The reconstruction results of in total 8 geometrical

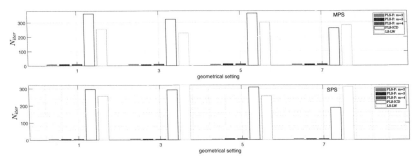

Figure 8.8: Computational cost for the three reconstruction algorithms for MPS *(first row)* and SPS *(second row)*.

	geometrical setting			
	G1	G3	G5	G7
MPS: FINESSE, $m = 2$	60	60	60	60
MPS: FINESSE, $m = 3$	160	128	192	160
MPS: FINESSE, $m = 4$	300	300	330	330
MPS: PLS-ICD	361	322	364	260
MPS: LS-LW	256	226	301	281
SPS: FINESSE, $m = 2$	30	36	24	36
SPS: FINESSE, $m = 4$	80	80	96	80
SPS: FINESSE, $m = 4$	150	120	180	150
SPS: PLS-ICD	296	292	308	187
SPS: LS-LW	256	321	256	311

Table 8.2: Total number of iterations for FINESSE, the ICD method, and the Landweber algorithm. The number N_{iter} was multiplied with the number of subspaces, $N_{iter} \cdot (2m^2 - 2)$, to allow for a more objective comparison.

settings were evaluated using quantitative basic metrics and a metric for computational cost. The results can be summarized as follows:

- A direct comparison of the reconstructed images against the ICD and Landweber reconstructions show that FINESSE returns reconstructed images that are very comparable in terms of image quality even though the short scan geometry reconstructions of FINESSE suffer from more streak artifacts (up to 80% for SPS) than the reconstructions obtained with the ICD method.

- The qualitative evaluation confirmed that there are differences in image quality that also depend on the pixel size. For MPS, the reconstruction error was always higher (between 5.0% and 18.6%) than with ICD or Landweber. However, the noise metrics showed that all three reconstruction methods perform very similarly. For SPS, FINESSE outperformed the Landweber algorithm in terms of reconstruction error by up to 50%. In terms of noise metrics, FINESSE outperformed the Landweber and the ICD method by up to 60%.

- An analysis of the computational cost showed that the Landweber and the ICD method need a very high number of iterates until convergence. Even if the number of sub-spaces of FINESSE is taken into account in the calculation, FINESSE outperforms both algorithms in 87.5% of the geometrical cases.

These results show that FINESSE is a real alternative reconstruction method in comparison to the ICD method or the Landweber algorithm. To avoid the presence of more short scan artifacts, another regularization function or other more aggressive regularization parameters may need to be applied.

8.4   Real Data Study

8.4.1   Measurement Setup

8.4.2   Scanner and Phantom Selection

FINESSE was also tested on real data. The measurement was done on the commercially available Siemens Healthineers CT scanner SOMATOM *Drive*. Table 8.3 lists the geometrical scan parameters of that CT scanner.

For the measurement, a conventional CT thorax/abdomen phantom was used since most of the scans that will be usually done on a CT scanner are done in the abdomen and thorax body regions. Figure 8.9 shows the top view topogram of the phantom that was used for the measurements. The topogram was derived from the measured data itself that means that the range displayed in the figure corresponds to the real measurement range of the scan.

For the scan, a standard abdomen protocol was selected. The scan was done at 120 kV tube voltage with 0.5s rotation time. The quantitative reference mAs of the scan was 210 mAs.

parameter description	parameter	value
focus-detector distance	R_{FD}	108.56 cm
focus-origin distance	R_F	59.5 cm
detector pixel size	Δu	0.1238 cm
detector pixel offset	u_{off}	1/4
number of detector pixels	N_u	736
number of projections per full turn	N_λ	1152

Table 8.3: Geometrical scan parameters of the commercially available Siemens Healthineers CT scanner SOMATOM Drive.

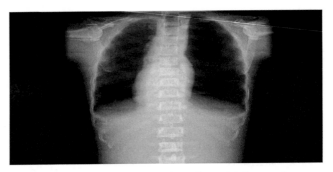

Figure 8.9: Top view topogram of the thorax/abdomen phantom.

8.4.2.1 FINESSE settings

The forward projection matrix $\mathbf{A}$ was formed using the distance-driven method. No statistical weighting matrix was applied during reconstruction, i.e., $\mathbf{W} = \mathbf{I}$. For each sub-iteration, σ_{max} and thereby L_s was estimated using the Power method [Golu 96]. The step size parameter t was always initialized with one. The algorithm was always initialized with a smooth FBP reconstruction.

The regularization term was formed using the quadratic potential with the parameter ϵ being 500. The regularization parameter applied in the reconstructions with real data was set to $\beta_R = 5$. The sub-iterations were stopped when the difference in the objective function became smaller than $T^{OF} = 10^{-5}$, with the embedded Newton-Raphson method being stopped when the change of the pixel value was less than $T^{NRM} = 10^{-6}$ HU.

Due to the results of the previous section, the block size was set to $m = 4$. The sub-spaces were sorted using the back-and-forth approach.

8.4.2.2   Reconstruction Methods

FINESSE was compared against the standard WFBP reconstruction method and against SAFIRE with strength 3. Both reconstruction methods are available on the CT scanner itself. WFBP is an analytical reconstruction method and SAFIRE is an iterative reconstruction method. All reconstructions were done on a image matrix size of 512×512 pixels. The slice width was set to 0.75 mm with a position increment of 0.75 mm. For the WFBP and SAFIRE reconstructions, the kernel Br40 was selected since it best matched with the resolution of the FINESSE reconstructions. The reconstruction FOV was 28cm.

8.4.3   Results

Figure 8.10 displays the reconstruction results of one arbitrary selected slice obtained with FINESSE, WFBP, and SAFIRE using both a standard abdomen and bone window. The slices were matched as well as possible; nevertheless, they do not match exactly. This is due to the reconstruction process. The WFBP and SAFIRE reconstructions were generated on the scanner, whereas the FINESSE images were generated in my personal C program. We show only one reconstructed image slice, since the others do not convey more information.

The reconstructions are very comparable in terms of image quality. The statistics in a circular ROI with radius of 16.3 mm in the left upper soft tissue region results in:

FINESSE: mean value $= 50.1$ HU, standard deviation: 21.49 HU

WFBP: mean value $= 50.1$ HU, standard deviation: 23.07 HU

SAFIRE: mean value $= 50.2$ HU, standard deviation: 17.69 HU

SAFIRE shows the lowest noise, followed by FINESSE and WFBP. The mean value is in all cases around 50 HU. The images were all free of artifacts.

8.4.4   Summary Discussion

In this section, we have presented reconstruction results of a thorax/abdomen phantom measured on a commercially available Siemens Healthineers CT scanner. The reconstruction result of FINESSE was compared against the analytical reconstruction method WFBP and the iterative reconstruction method SAFIRE of strength 3, which were both available on the CT scanner. The reconstruction results were evaluated by visual inspection and by using the statistics of a circular ROI in the phantom. The results can be summarized as follows:

- A visual inspection of the reconstructed slices showed no artifacts or other conspicuous features.

- The quantitative evaluation of the circular ROI confirmed the visual differences in terms of noise: SAFIRE showed lowest noise, followed by FINESSE and WFBB, but it is important to note that FINESSE was applied with a quadratic regularizer whereas SAFIRE uses edge-preserving regularization.

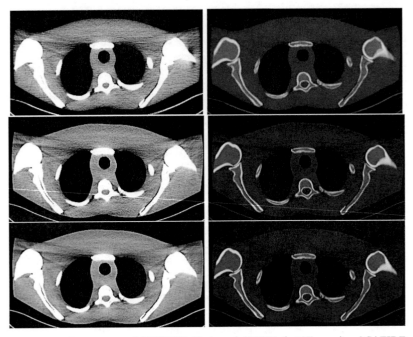

Figure 8.10: Reconstructed FINESSE *(first row)*, WFBP *(middle row)* and SAFIRE *(bottom row)* images displayed using a standard abdomen window (c/w=40/300) *(left column)* and bone window (c/w=480/2500) *(right column)*.

Altogether, FINESSE also works using real data. It is probably possible to even reduce the image noise in the FINESSE reconstruction by adapting the free parameters of the reconstruction algorithm, such as regularization parameter β_R and/or the regularization function.

8.5 Discussion and Conclusion

With the development of our novel statistical iterative reconstruction, called FI-NESSE, we want to overcome most of the difficulties that are related to iterative reconstruction methods. Often iterative reconstruction methods could not satisfy all or most of the desired properties of fast and global convergence, of being stable and convergent, and of being fast, robust, parallelizable, and flexible. By integrating the principles of FISTA and block-based coordinate descent methods, we retained the advantages of each method while overcoming their deficiencies. The description of the algorithm design in the first part of this chapter involves a presentation of the general concept and the mathematical formulation of FINESSE. As a result, it appears that FINESSE provides a non-linear exact sub-space search. Moreover, the image denoising problem can be solved in parallel.

Our simulation study included a large variety of scanning geometries and a fair number of repeated scans in each geometry to assess dependence on photon statistics. The geometries were representative of contemporary CT scans. Two different image pixel sizes and a challenging phantom, namely the FORBILD head phantom, were used. Moreover, the FINESSE specific block size parameter was varied: $m = 2, 3, 4$. FINESSE reconstructions were compared against resolution matched Landweber and ICD reconstruction which made a thorough investigation of FINESSE possible. A visual inspection of the noiseless reconstructions showed first differences in terms of image quality. Whereas the ICD method suppresses most short scan artifacts such as streaks in the image, FINESSE could not handle these as well as ICD, but did much better than the Landweber algorithm. On the other hand, the full scan reconstructions do not show obvious differences in terms of image quality.

The evaluation in the simulation study also involved basic metrics covering discretization errors, image pixel noise, statistical correlation between pixels, and computational cost. The bias was evaluated in both the full background of the FORBILD head phantom and in an elliptical region around the ear. For MPS, the bias was always in all geometries higher (between 5.0% and 18.6%) than for ICD and Landweber, whereas, for SPS, the bias of FINESSE was between ICD and Landweber. Moreover, the bias measurement dependents little on the block size. In all geometries, the block size $m = 4$ resulted in the lowest bias. The noise in the image was suppressed best with FINESSE in the MPS short scan geometries and best with Landweber in the full scan geometries. For SPS, the noise measure was always lowest using FINESSE. All algorithms have correlation coefficients with the same sign, meaning pixels are correlated in the same direction. That measure shows no clear trend, sometimes FINESSE has the lowest correlation coefficients, while, in other geometries, ICD or Landweber show the lowest numbers. In terms of computational cost, FINESSE clearly outperforms the Landweber and the ICD method.

A thorax/abdomen phantom was scanned on a commercially available Siemens Healthineers CT scanner SOMATOM Drive. The data was reconstructed using the analytical reconstruction method WFBP and the iterative reconstruction method SAFIRE both available on the scanner. These reconstructions were compared against offline reconstructed FINESSE images. In none of the reconstructions were artifacts present. They all are of comparable image quality with no obvious differences in the images. An evaluation of a circular ROI showed that the images have the same mean value within the ROI, whereas the standard deviation of the SAFIRE reconstruction was lowest followed by the FINESSE reconstruction.

Both studies show that FINESSE is stable and globally convergent, although a mathematical proof of convergence is very complicated. Furthermore, the algorithm is robust, parallelizable, and somehow flexible in terms of the block size parameter m and the regularization function, providing the function is convex. Altogether, FINESSE is a real alternative to other iterative reconstruction methods.

Summary and Outlook

9.1 Summary

The topics discussed in this dissertation cover a thorough analysis of the impact of various influencing factors in iterative reconstruction algorithms as well as a novel iterative reconstruction method and are summarized below.

The introduction of this dissertation in Chapter 1 discusses the popularity and advantages of CT in medical imaging as well as the drawback of adding additional radiation exposure to the patient. In the past, these pros and cons of CT examinations had opened a wide field of research towards low-dose CT, also including statistical iterative reconstruction algorithms. The chapter concludes with a summary of the scientific contribution of the presented work and the organization of the dissertation.

Chapter 2 gave a brief introduction in the basic components of a computed tomography scanner. That was followed by a description of important gold standard techniques such as standard scan modes, beam filtration, quarter detector offset, and flying focal spot.

The mathematical description of the measurement process together with the definition of several coordinate systems was given in Chapter 3. Moreover, the mathematical formulas of the Radon transform for the parallel-beam and fan-beam geometry were defined in that chapter.

The basics of iterative reconstruction techniques, the description of the definition of the objective function, and the definition of a non-constrained and constrained image reconstruction problem were explained in Chapter 4. Furthermore, the mathematical formulation of two commonly known iterative reconstruction methods, namely the Landweber method and the ICD method, was given. The chapter concluded with a discussion of the strength and weaknesses of iterative reconstruction algorithms.

In Chapter 5, a complete and detailed framework for the image quality assessment used throughout the dissertation has been presented. Any change in the preprocessing of the measured data or in the reconstruction process could have influence on the final IQ. For doctors, it is very important that the final reconstructed image is related to how well it conveys all anatomical or functional information. The signs of disease or injury need to be clearly visible such that the interpreting radiologist can make an accurate diagnosis. Therefore, the simulation studies in this disserta- tion were based on a mathematical phantom, the FORBILD head phantom. With the knowledge of the exact definition of the phantom, a thorough quantitative anal- ysis using basic and task-based metrics was possible. The quantitative evaluation using basic metrics involved spatial resolution, computational cost, reconstruction error, and noise. For the task-based IQ assessment, the ideal observer, Channelized Hotelling observer, and human observer were used and described in that chapter.

Chapter 6 discussed the impact of discrete image representation techniques on IQ. Three commonly used discrete image representation techniques were presented: i) the sampling approach including Joseph's method; ii) the strip integral approach including the distance-driven method; and, iii) the basis function approach including the B-splines and the blobs. A first understanding of how forward projection models impact IQ was obtained in a preliminary study that covered altogether seven image representation techniques (Joseph's method, distance-driven method, B-splines of order $n = 0, 1, 2, 3$ and blobs). Based on the results of a qualitative and quantitative evaluation, the methods could be divided into three groups. Within each group the differences in basic metrics were small whereas across the groups the difference was more significant. Due to the preliminary study results, an extensive follow- up study was set up with linear interpolation models, namely with the Joseph's method, the distance-driven method, and the bilinear method (B-splines of order $n = 1$). That study covered several geometrical settings with noiseless and noisy reconstructions. The quantitative evaluation using basic metrics clearly showed that there are important differences in image characteristics. Especially when using a small pixel size, the differences turned out to be more significant. However, the observed differences were not such that one method could be identified as favored over all others in all settings. Therefore, an extensive quantitative evaluation using task-based metrics was set up. Small differences in the AUC measure were observed. However, in all cases these differences were not statistically significant. In summary, each forward projection model showed its own characteristics in terms of basic metrics, whereas these differences totally balanced out when performing a task-based assessment.

Next, the impact of the regularization method on IQ in a statistical iterative re- construction problem was evaluated in Chapter 7. Two regularization methods were considered: i) a regularization based on stopping the iterative reconstruction algo- rithm after a finite number of iteration steps; and, ii) a regularization that is enforced directly by a constraint in the definition of the objective function. The solution of the first regularization method was found by applying 1000 iteration steps using the Landweber algorithm. The solution of the second regularization method was the out- put of the ICD algorithm, when the change of each pixel value in the reconstructed volume from one to the next iteration became smaller than $10^{-5}/cm$. The regular- ization term of the ICD algorithm was formed using both the Fair potential and the quadratic potential. The investigation of the regularization method was subdivided into two parts each covering a different definition of the statistical weighting matrix. In both parts, IQ was assessed using basic metrics. In the first part, the statistical weighting matrix was represented by non-uniform weights related to a bowtie filter in the beam path. Two shapes of a bowtie filter were taken into account and compared against reconstruction results with no application of a statistical weighting matrix. It was shown that the presence of a statistical weighting matrix may easily introduce significant errors that differ from resolution errors. While a regularization based on stopping after a given number of iterations is more sensitive to a lack of smoothness in the statistical weighing matrix, a regularization that is directly enforced by a constraint seems to be much more stable in terms of reconstruction artifacts. Moreover, the presence of a statistical weighting matrix increased the reconstruction error, while decreasing the noise measure in both regularization methods. In the second part, the statistical weighting matrix carried the information about redundant data. Two definitions for the data redundancy handling were investigated and compared against reconstruction results with no application of a statistical weighting matrix. The study showed that the use of a data redundancy handling weight reduces reconstruction artifacts and thereby decrease the reconstruction error. Moreover, the mean standard deviation decreased while increasing correlation between pixels for both regularization methods. Both study parts showed that one cannot straightforwardly state that a weighting matrix always helps to reduce reconstruction errors even though it models the physics better. Moreover, the noise-resolution trade-off introduced by the iteration number is not always attractive for CT imaging. In such situations, the penalized (weighted) maximum likelihood solution is a more attractive procedure.

A novel iterative reconstruction algorithm for CT imaging, called FINESSE, was introduced in Chapter 8. The algorithm integrates together the principles of FISTA and block-based coordinate descent methods. A detailed description of the algorithm design was given also including information about the FISTA algorithm. Next, FINESSE was compared against the ICD method and the Landweber algorithm in a simulation study. The qualitative evaluation using basic metrics confirmed that there are differences in IQ between the investigated algorithms, but these results were not such that one of the algorithms could be said to be preferred. The chapter concludes with a real data study comparing the results of FINESSE, WFBP, and SAFIRE of strength 3.

9.2   Outlook

This dissertation provides a detailed analysis of the influence of various parameter settings in statistical iterative reconstruction methods on image quality. Apparently, statistical iterative reconstruction methods have, per default, many freely selectable parameters. Currently, there is still a huge lack of thorough investigation in the direction of how which parameter setting affects the final image quality. To name but a few examples, the influence of a positivity constraint in the objective function or the design of the regularization term are topics that are controversial and widely discussed. Yet, exactly such fundamental studies are essential to attain a better understanding of how to gain the most out of statistical iterative reconstruction methods.

To set up an extensive investigation is not trivial. Many items must be considered beforehand since any of them might influence the total study setup. Some questions that need to be answered are the following: What do we want to prove/show? Which influence does the parameter might have on the final image quality? Do we need reconstructions of various reconstruction algorithms, such as finding the solution of a non-constrained and constrained image reconstruction problem? Do we expect different results depending on the reconstructed image pixel size? What are representative geometrical CT settings? What phantom should be used? How do we assess the final image quality? These are just some of the questions that arise in this context. Depending on the output of the assessment, even more questions might arise that may lead to an extension of the original study design. All in all, the investigation of single parameters is usually quite time consuming. On the one hand, quite a lot of programming effort might be needed beforehand if certain programs are not yet available, and, on the other hand, generating all the data output often leads to a huge computational effort. Nevertheless, such studies are very essential and valuable even though they are often unpopular. However, precisely because the demand for deep learning methods is increasing, such investigations are becoming increasingly important. Deep learning algorithms rely on statistical iterative reconstruction techniques, e.g., Canon offers a network to obtain the MBIR result faster, and others include a finite number of iterations within a network with trainable regularizers. The success of all these still depends on understanding the impact of parameters that were investigated throughout this dissertation.

While statistical iterative reconstruction methods are just one aspect to reduce the patient's radiation dose, hardware development is another important step in that direction. This includes, in particular, the data measurement system, i.e., the amount of electronic noise of the detector itself. Beside modern energy integrating detectors that have little electronic noise, photon counting detectors are said to be free of electronic noise. Here, the question arises, whether and to what degree that might have influence on the design in the statistical iterative reconstruction algorithms.

Proofs and Derivations

A.1 Proofs for the Landweber algorithm

This section provides both the proof of the convergence fixed point and of the limits of the controlling parameter η for the Landweber algorithm, for which the formal description of the iteration schema,

$$\boldsymbol{f}^{(n+1)} = \boldsymbol{f}^{(n)} + \eta \, \tilde{\mathbf{A}}^T \left(\tilde{\boldsymbol{g}} - \tilde{\mathbf{A}} \boldsymbol{f}^{(n)} \right) \, , \tag{A.1}$$

was given in Section 4.3.1.

Let η be a fixed parameter that does not change within the iterations and let

$$\boldsymbol{\Delta} = \mathbf{I} - \eta \tilde{\mathbf{A}}^T \tilde{\mathbf{A}} \, . \tag{A.2}$$

Then, Equation A.1 can be rewritten as follows

$$
\begin{aligned}
\boldsymbol{f}^{(n+1)} &= \boldsymbol{\Delta} \boldsymbol{f}^{(n)} + \eta \tilde{\mathbf{A}}^T \tilde{\boldsymbol{g}} \\
&= \boldsymbol{\Delta}^2 \boldsymbol{f}^{(n-1)} + (\boldsymbol{\Delta} + \mathbf{I}) \eta \tilde{\mathbf{A}}^T \tilde{\boldsymbol{g}} \\
&= \dots \\
&= \boldsymbol{\Delta}^{(n+1)} \boldsymbol{f}^{(0)} + (\boldsymbol{\Delta}^{(n)} + \dots + \boldsymbol{\Delta} + \mathbf{I}) \eta \tilde{\mathbf{A}}^T \tilde{\boldsymbol{g}} \\
&= \boldsymbol{\Delta}^{(n+1)} \boldsymbol{f}^{(0)} + \eta \sum_{l=0}^{n} \boldsymbol{\Delta}^{(l)} \tilde{\mathbf{A}}^T \tilde{\boldsymbol{g}} \, .
\end{aligned}
\tag{A.3}
$$

Suppose that $\boldsymbol{f}^{(0)} = 0$ and let's assume that the maximum singular value of $\boldsymbol{\Delta}$ is smaller than one, then the first term in Equation A.3 is equal to zero and the second term, which represents the Neumann series, converge to $(\mathbf{I} - \boldsymbol{\Delta})^{-1}$. Thus, the Landweber algorithm converge to the following fixed point

$$
\begin{aligned}
\boldsymbol{f}^* = \lim_{n \to \infty} \boldsymbol{f}^{(n+1)} &= (\mathbf{I} - \boldsymbol{\Delta})^{-1} \eta \tilde{\mathbf{A}}^T \tilde{\boldsymbol{g}} \\
&= \left(\tilde{\mathbf{A}}^T \tilde{\mathbf{A}} \right)^{-1} \tilde{\mathbf{A}}^T \tilde{\boldsymbol{g}} \, .
\end{aligned}
\tag{A.4}
$$

However, the convergence to the fixed point is only guaranteed if and only if the parameter η lies within a certain interval. In order to prove this, we can use the singular value decomposition (SVD) of $\tilde{\mathbf{A}}$ that is $\tilde{\mathbf{A}} = \mathbf{U} \boldsymbol{\Sigma} \mathbf{V}^T$ with $\mathbf{U} \mathbf{U}^T = \mathbf{U}^T \mathbf{U} = \mathbf{I}$,

$\mathbf{V}\mathbf{V}^T = \mathbf{V}^T\mathbf{V} = \mathbf{I}$, and $\boldsymbol{\Sigma}$ being a diagonal matrix with the singular eigenvalues, σ_i, of $\tilde{\mathbf{A}}$ on its diagonal. Using the SVD for $\tilde{\mathbf{A}}$ in Equation A.2 yields

$$\boldsymbol{\Delta} = \mathbf{I} - \eta\mathbf{V}\boldsymbol{\Sigma}^T\mathbf{U}^T\mathbf{U}\boldsymbol{\Sigma}\mathbf{V}^T$$
$$= \mathbf{V}\left(\mathbf{I} - \eta\boldsymbol{\Sigma}^T\boldsymbol{\Sigma}\right)\mathbf{V}^T . \tag{A.5}$$

Next, $\boldsymbol{\Delta}$ and $\tilde{\mathbf{A}}$ in Equation A.3 is substituted by Equation A.5 and the SVD, respectively. Furthermore, suppose again that $\boldsymbol{f}^{(0)} = 0$. Then,

$$\lim_{n\to\infty} \boldsymbol{f}^{(n+1)} = \eta\mathbf{V}\sum_{l=0}^{\infty}\left(\mathbf{I} - \eta\boldsymbol{\Sigma}^T\boldsymbol{\Sigma}\right)^l \mathbf{V}^T\mathbf{V}\boldsymbol{\Sigma}^T\mathbf{U}^T\tilde{\boldsymbol{g}}$$
$$= \eta\mathbf{V}\sum_{l=0}^{\infty}\left(\mathbf{I} - \eta\boldsymbol{\Sigma}^T\boldsymbol{\Sigma}\right)^l \boldsymbol{\Sigma}^T\mathbf{U}^T\tilde{\boldsymbol{g}} . \tag{A.6}$$

The equation above represents a Neumann series, which only converge if and only if $(\mathbf{I} - \eta\boldsymbol{\Sigma}^T\boldsymbol{\Sigma})$ is smaller than one. Hence, the series for each diagonal element converges if $|1 - \eta\sigma_i^2| < 1$, i.e.,

$$-1 < (1 - \eta\sigma_i^2) < 1$$
$$\Leftrightarrow \qquad 0 < \eta < \frac{2}{\sigma_i^2} , \tag{A.7}$$

which is guaranteed if $0 < \eta < \frac{2}{\sigma_{\max}^2}$, where $\sigma_{\max}$ is the maximum eigenvalue of $\tilde{\mathbf{A}}$.

List of Abbreviations and Symbols

Abbreviations

Symbols

Scalars: geometry and sampling

Scalars: physical quantities

Scalars: valued functions

Vectors and matrices

Transforms

Iterative reconstruction parameters, vectors and matrices

Scalars: image quality evaluation

Discrete image representation techniques: parameters and valued functions

Data simulation parameters

Weighting matrix: parameters and functions

FINESSE: scalars, parameters and functions

List of Figures

List of Tables

List of Algorithms

Bibliography

[Abat 82]   T. Abatzoglou and B. O'Donnell. "Minimization by coordinate descent". *Journal of Optimization Theory and Applications*, Vol. 36, No. 2, pp. 163–174, 1982.

[Abbe 01]   C. K. Abbey and H. H. Barrett. "Human-and model-observer performance in ramp-spectrum noise: effects of regularization and object variability". *JOSA A*, Vol. 18, No. 3, pp. 473–488, 2001.

[Barr 04]   H. H. Barrett and K. J. Myers. *Foundations of Image Sience*. Wiley, 2004.

[Barr 12]   H. H. Barrett and W. Swindell. *Radiological imaging: the theory of image formation, detection, and processing*. Elsevier, 2012.

[Bays 12]   H. Baysson, C. Etard, H. Brisse, and M. Bernier. "Diagnostic radiation exposure in children and cancer risk: current knowledge and perspectives". *Archives de pediatrie: organe officiel de la Societe francaise de pediatrie*, Vol. 19, No. 1, pp. 64–73, 2012.

[Beck 09]   A. Beck and M. Teboulle. "A Fast Iterative Shrinkage-Thresholding Algorithm for Linear Inverse Problems". *SIAM J. Imaging Sciences*, Vol. 2, No. 1, pp. 183–202, 2009.

[Bens 10]   T. M. Benson, B. K. De Man, L. Fu, and J.-B. Thibault. "Block-based iterative coordinate descent". In: *Nuclear Science Symposium Conference Record (NSS/MIC), 2010 IEEE*, pp. 2856–2859, IEEE, 2010.

[Boum 93]   C. Bouman and K. Sauer. "A generalized Gaussian image model for edge-preserving MAP estimation". *Image Processing, IEEE Transactions on*, Vol. 2, No. 3, pp. 296–310, 1993.

[Boum 96]   C. Bouman, K. Sauer, *et al.* "A unified approach to statistical tomography using coordinate descent optimization". *Image Processing, IEEE Transactions on*, Vol. 5, No. 3, pp. 480–492, 1996.

[Bows 98]   J. Bowsher, M. Smith, J. Peter, and R. Jaszczak. "A comparison of OSEM and ICD for iterative reconstruction of SPECT brain images". In: *Journal of Nuclear Medicine*, pp. 79P–79P, SOC NUCLEAR MEDICINE INC 1850 SAMUEL MORSE DR, RESTON, VA 20190-5316 USA, 1998.

[Bren 07]   D. J. Brenner and E. J. Hall. "Computed tomography - an increasing source of radiation exposure". *New England Journal of Medicine*, Vol. 357, No. 22, pp. 2277–2284, 2007.

[Bund 20]   Bundesamt für Strahlenschutz. "X-ray diagnostics: Frequency and radiation exposure". Feb. 2020.

[Burg 88]   A. Burgess and B. Colborne. "Visual signal detection. IV. Observer inconsistency". *JOSA A*, Vol. 5, No. 4, pp. 617–627, 1988.

[Bush 11]   J. T. Bushberg and J. M. Boone. *The essential physics of medical imaging*. Lippincott Williams & Wilkins, 2011.

[Buzu 04]   T. M. Buzug. *Einführung in die Computertomographie: Mathematisch-physikalische Grundlagen der Bildrekonstruktion.* Springer-Verlag, Berlin Heidelberg, 2004.

[Buzu 08]   T. M. Buzug. *Computed Tomography: From Photon Statistics to Modern Cone-Beam CT.* Springer-Verlag, Berlin Heidelberg, 2008.

[Case 01]   G. Casella and R. L. Berger. *Statistical Inference.* Vol. 2, Duxbury Pacific Grove, CA, 2001.

[Chen 06]   G.-H. Chen, R. Tokalkanahalli, T. Zhuang, B. E. Nett, and J. Hsieh. "Development and evaluation of an exact fan-beam reconstruction algorithm using an equal weighting scheme via locally compensated filtered backprojection (LCFBP)". *Medical physics*, Vol. 33, No. 2, pp. 475–481, 2006.

[Cutt 09]   J. M. Cuttler and M. Pollycove. "Nuclear energy and health: and the benefits of low-dose radiation hormesis". *Dose-Response*, Vol. 7, No. 1, pp. dose–response, 2009.

[Dani 03]   P.-E. Danielsson and M. Magnusson Seger. "A Proposal for Combining FBP and ART in CT-reconstruction". In: *7th International Meeting on Fully Three-Dimensional Image Reconstruction in Radiology and Nuclear Medicine*, 2003.

[Daub 04]   I. Daubechies, M. Defrise, and C. D. Mol. "An iterative thresholding algorithm for linear inverse problems with a sparsity constraint". *Comm. Pure Appl. Math.*, Vol. 57, pp. 1413–57, 2004.

[De G 09]   A. B. De González, M. Mahesh, K.-P. Kim, M. Bhargavan, R. Lewis, F. Mettler, and C. Land. "Projected cancer risks from computed tomographic scans performed in the United States in 2007". *Archives of internal medicine*, Vol. 169, No. 22, pp. 2071–2077, 2009.

[De M 00]   B. De Man, J. Nuyts, P. Dupont, G. Marchal, and P. Suetens. "Reduction of metal streak artifacts in x-ray computed tomography using a transmission maximum a posteriori algorithm". *IEEE transactions on nuclear science*, Vol. 47, No. 3, pp. 977–981, 2000.

[De M 01]   B. De Man, J. Nuyts, P. Dupont, G. Marchal, and P. Suetens. "An iterative maximum-likelihood polychromatic algorithm for CT". *Medical Imaging, IEEE Transactions on*, Vol. 20, No. 10, pp. 999–1008, 2001.

[De M 02]   B. De Man and S. Basu. "Distance-driven projection and backprojection". In: *Nuclear Science Symposium Conference Record, 2002 IEEE*, pp. 1477–1480, IEEE, 2002.

[Desa 12]   G. S. Desai, R. N. Uppot, E. W. Yu, A. R. Kambadakone, and D. V. Sahani. "Impact of iterative reconstruction on image quality and radiation dose in multidetector CT of large body size adults". *European Radiology*, Vol. 22, No. 8, pp. 1631–1640, 2012.

[Doss 13]   O. Dössel. *Bildgebende Verfahren in der Medizin: von der Technik zur medizinischen Anwendung.* Springer-Verlag, 2013.

[Eber 11]   D. Eberly. "Distance from a Point to an Ellipse, an Ellipsoid, or a Hyperellipsoid". *Geometric Tools, LLC*, 2011.

[Ecks 03]   M. Eckstein, J. Bartroff, C. Abbey, J. Whiting, and F. Bochud. "Automated computer evaluation and optimization of image compression of x-ray coronary angiograms for signal known exactly detection tasks". *Optics Express*, Vol. 11, No. 5, pp. 460–475, 2003.

[Elba 02]   I. Elbakri, J. Fessler, *et al.* "Statistical image reconstruction for polyenergetic X-ray computed tomography". *Medical Imaging, IEEE Transactions on*, Vol. 21, No. 2, pp. 89–99, 2002.

[Erdo 99]   H. Erdogan and J. A. Fessler. "Ordered subsets algorithms for transmission tomography". *Physics in medicine and biology*, Vol. 44, No. 11, p. 2835, 1999.

[Expe 14]   "Expert opinion: Are CT scans safe?". Sciencedaily, 2014.

[Fahi 13]   B. P. Fahimian, Y. Zhao, Z. Huang, R. Fung, Y. Mao, C. Zhu, M. Khatonabadi, J. J. DeMarco, s. J. Osher, M. F. McNitt-Gray, and J. Miao. "Radiation dose reduction in medical x-ray CT via Fourier-based iterative reconstruction". *Med. Phys.*, Vol. 40, No. 3, p. 031914, 2013.

[Fair 74]   R. C. Fair. "On the robust estimation of econometric models". In: *Annals of Economic and Social Measurement, Volume 3, number 4*, pp. 667–677, NBER, 1974.

[Feld 84]   L. A. Feldkamp, L. C. David, and J. W. Kress. "Practical cone-beam algorithm". *Journal of Optical Society of America*, Vol. 1, No. 6, pp. 612–619, 1984.

[Fess 11]   J. A. Fessler and D. Kim. "Axial block coordinate descent (ABCD) algorithm for X-ray CT image reconstruction". *Proc. of the 11th international meeting on Fully 3D image reconstruction in radiology and nuclear medicine*, pp. 262–5, 2011.

[Fess 94]   J. A. Fessler. "Penalized Weighted Least-Squares Image Reconstruction for Positron Emission Tomography". *IEEE Transaction on Medical Imaging*, Vol. 13, No. 2, pp. 290–300, 1994.

[Fess 97]   J. Fessler, E. P. Ficaro, N. H. Clinthorne, K. Lange, *et al.* "Grouped-coordinate ascent algorithms for penalized-likelihood transmission image reconstruction". *Medical Imaging, IEEE Transactions on*, Vol. 16, No. 2, pp. 166–175, 1997.

[Fess 99]   J. A. Fessler and S. D. Booth. "Conjugate-gradient preconditioning methods for shift-variant PET image reconstruction". *IEEE Trans. Im. Proc.*, Vol. 8, No. 5, pp. 688–99, 1999.

[Floh 03]   T. Flohr, K. Stierstorfer, H. Bruder, J. Simon, A. Polacin, and S. Schaller. "Image reconstruction and image quality evaluation for a 16-slice CT scanner". *Medical physics*, Vol. 30, No. 5, pp. 832–845, 2003.

[Floh 05]   T. Flohr, K. Stierstorfer, S. Ulzheimer, H. Bruder, A. Primak, and C. Mc-Collough. "Image reconstruction and image quality evaluation for a 64-slice CT scanner with z-flying focal spot". *Medical physics*, Vol. 32, No. 8, pp. 2536–2547, 2005.

[Food 17]   Food and D. Administration. "What are the Radiation Risks from CT?". Food and Drug Administration, May 2017.

[Furl 10]   B. Furlow. "Radiation dose in computed tomography". *Radiologic Technology*, Vol. 81, No. 5, pp. 437–450, 2010.

[Geng 10]   L. Z. Gengsheng. *Medical Image Reconstruction: A Conceptual Tutorial.* Higher Education Press, 2010.

[Giff 07]   H. Gifford, P. Kinahan, C. Lartizien, and M. King. "Evaluation of multiclass model observers in PET LROC studies". *Nuclear Science, IEEE Transactions on*, Vol. 54, No. 1, pp. 116–123, 2007.

[Gold 07]   L. W. Goldman. "Principles of CT and CT technology". *Journal of nuclear medicine technology*, Vol. 35, No. 3, pp. 115–128, 2007.

[Golu 96]   G. H. Golub and C. F. van Loan. *Matrix Computations.* John Hopkins University Press, 1996.

[Hahn 15a]   K. Hahn, U. Rassner, H. Davidson, H. Schöndube, K. Stierstorfer, J. Hornegger, and F. Noo. "Iterative CT reconstruction with small pixel size: distance-driven forward projector versus Joseph's". In: *SPIE Medical Imaging*, pp. 94123D–94123D, International Society for Optics and Photonics, 2015.

[Hahn 15b]   K. Hahn, H. Schöndube, K. Stierstorfer, and Noo. "Impact of statistical weights and edge preserving regularization on image quality in iterative CT reconstruction". In: M. King, S. Glick, and K. Mueller, Eds., *Fully Three-Dimensional Image Reconstruction in Radiology and Nuclear Medicine*, pp. 51–54, Newport, Rhode Island, USA, May 31 - June 4 2015.

[Hahn 16]   K. Hahn, H. Schöndube, K. Stierstorfer, J. Hornegger, and F. Noo. "A comparison of linear interpolation models for iterative CT reconstruction". *Medical physics*, Vol. 43, No. 12, pp. 6455–6473, 2016.

[Hans 98]   P. C. Hansen. *Rank-deficient and discrete ill-posed problems: numerical aspects of linear inversion.* Vol. 4, Siam, 1998.

[Hara 09]   A. K. Hara, R. G. Paden, A. C. Silva, J. L. Kujak, H. J. Lawder, and W. Pavlicek. "Iterative Reconstruction Technique for Reducing Body Radiation Dose at CT: Feasibility Study". *AJR*, Vol. 193, pp. 764–771, 2009.

[Hebe 88]   T. Hebert, R. Leahy, and M. Singh. "Fast MLE for SPECT using an intermediate polar representation and a stopping criterion". *Nuclear Science, IEEE Transactions on*, Vol. 35, No. 1, pp. 615–619, 1988.

[Herm 09]   G. T. Herman. *Fundamentals of computerized tomography: image reconstruction from projections.* Springer, 2009.

[hila 11]   hil/aerzteblatt.de. "MRT laut Barmer Arztreport in Deutschland am haeufigsten". *aerzteblatt.de*, 2011.

[Hofm 14]   C. Hofmann, M. Knaup, and M. Kachelrieß. "Effects of ray profile modeling on resolution recovery in clinical CT". *Medical physics*, Vol. 41, No. 2, p. 021907, 2014.

[Horb 02]   S. Horbelt, M. Liebling, and M. Unser. "Discretization of the Radon transform and of its inverse by spline convolutions". *Medical Imaging, IEEE Transactions on*, Vol. 21, No. 4, pp. 363–376, 2002.

[Hsie 09]   J. Hsieh. *Computed tomography: principles, design, artifacts, and recent advances.* SPIE - The International Society for Optical Engineering, Bellingham, Washington, USA, 2009.

[Hubb 04]   J. H. Hubbell and s. M. Seltzer. "Tables of X-Ray Mass Attenuation Coefficients and Mass Energy-Absorption Coefficients from 1 keV to 20 MeV for Elements Z = 1 to 92 and 48 Additional Substances of Dosimetric Interest". 2004.

[Hube 11]   P. J. Huber. *Robust statistics.* Springer, 2011.

[Huds 94]   H. Hudson and R. Larkin. "Accelerated image reconstruction using ordered subsets of projection data". *IEEE Transaction on Medical Imaging*, Vol. 13, No. 4, pp. 601–609, 1994.

[Jian 03]   M. Jiang and G. Wang. "Convergence studies on iterative algorithms for image reconstruction". *Medical Imaging, IEEE Transactions on*, Vol. 22, No. 5, pp. 569–579, 2003.

[Jose 82]   P. M. Joseph. "An improved algorithm for reprojecting rays through pixel images". *Medical Imaging, IEEE Transactions on*, Vol. 1, No. 3, pp. 192–196, 1982.

[Judy 76]   P. Judy. "The line spread function and modulation transfer function of a computed tomographic scanner". *Medical physics*, Vol. 3, No. 4, pp. 233–236, 1976.

[Kadr 09]   D. J. Kadrmas, M. E. Casey, N. F. Black, J. J. Hamill, V. Y. Panin, and M. Conti. "Experimental comparison of lesion detectability for four fully-3D PET reconstruction schemes". *Medical Imaging, IEEE Transactions on*, Vol. 28, No. 4, pp. 523–534, 2009.

[Kak 01]   A. C. Kak and M. Slaney. *Principles of computerized tomographic imaging.* Society for Industrial and Applied Mathematics, http://www.slaney.org/pct/pct-toc.html, 2001.

[Kale 06]   W. A. Kalender. *Computertomographie: Grundlagen, Gerätetechnologie, Bildqualität, Anwendungen.* Publicis MCD Verlag, München, 2nd Ed., 2006.

[Kamp 98]   C. Kamphuis and F. J. Beekman. "Accelerated iterative transmission CT reconstruction using an ordered subsets convex algorithm". *Medical Imaging, IEEE Transactions on*, Vol. 17, No. 6, pp. 1101–1105, 1998.

[Kats 03]   A. Katsevich. "A general scheme for constructing inversion algorithms for cone beam CT". *International Journal of Mathematics and Mathematical Sciences*, Vol. 2003, No. 21, pp. 1305–1321, 2003.

[Kim 15]   D. Kim, S. Ramani, and J. A. Fessler. "Combining ordered subsets and momentum for accelerated X-ray CT image reconstruction". *IEEE transactions on medical imaging*, Vol. 34, No. 1, pp. 167–178, 2015.

[Land 51]   L. Landweber. "An iteration formula for Fredholm integral equations of the first kind". *American journal of mathematics*, pp. 615–624, 1951.

[LaRo 07]   S. J. LaRoque, E. Y. Sidky, D. C. Edwards, and X. Pan. "Evaluation of the channelized Hotelling observer for signal detection in 2D tomographic imaging". In: *Medical Imaging*, pp. 651514–651514, International Society for Optics and Photonics, 2007.

[Laur]   G. Lauritsch and H. Bruder. "FORBILD head phantom". online, last
            accessed O 7, 2014.

[Lewi 90]   R. M. Lewitt. "Multidimensional digital image representations using
            generalized Kaiser–Bessel window functions". *JOSA A*, Vol. 7, No. 10,
            pp. 1834–1846, 1990.

[Lewi 92]   R. M. Lewitt. "Alternatives to voxels for image representation in iterative
            reconstruction algorithms". *Physics in Medicine and Biology*, Vol. 37,
            No. 3, p. 705, 1992.

[Li 07]   B. Li, G. B. Avinash, and J. Hsieh. "Resolution and noise trade-off
            analysis for volumetric CT". *Medical physics*, Vol. 34, No. 10, pp. 3732–
            3738, 2007.

[Llac 89]   J. Llacer and E. Veklerov. "Feasible images and practical stopping rules
            for iterative algorithms in emission tomography". *Medical Imaging, IEEE
            Transactions on*, Vol. 8, No. 2, pp. 186–193, 1989.

[Luo 92]   Z.-Q. Luo and P. Tseng. "On the convergence of the coordinate descent
            method for convex differentiable minimization". *Journal of Optimization
            Theory and Applications*, Vol. 72, No. 1, pp. 7–35, 1992.

[Mate 96]   S. Matej and R. M. Lewitt. "Practical considerations for 3-D image
            reconstruction using spherically symmetric volume elements". *Medical
            Imaging, IEEE Transactions on*, Vol. 15, No. 1, pp. 68–78, 1996.

[Math 13]   J. D. Mathews, A. V. Forsythe, Z. Brady, M. W. Butler, S. K. Goergen,
            G. B. Byrnes, G. G. Giles, A. B. Wallace, P. R. Anderson, T. A. Guiver,
            et al. "Cancer risk in 680 000 people exposed to computed tomogra-
            phy scans in childhood or adolescence: data linkage study of 11 million
            Australians". *Bmj*, Vol. 346, p. f2360, 2013.

[McCo 15]   C. H. McCollough, J. T. Bushberg, J. G. Fletcher, and L. J. Eckel.
            "Answers to common questions about the use and safety of CT scans".
            In: *Mayo Clinic Proceedings*, pp. 1380–1392, Elsevier, 2015.

[Metz 08]   C. E. Metz. "ROC analysis in medical imaging: a tutorial review of the
            literature". *Radiological physics and technology*, Vol. 1, No. 1, pp. 2–12,
            2008.

[Metz 78]   C. E. Metz. "Basic principles of ROC analysis". In: *Seminars in nuclear
            medicine*, pp. 283–298, Elsevier, 1978.

[Myer 00]   K. J. Myers. "Ideal observer models of visual signal detection". *Handbook
            of medical imaging*, Vol. 1, pp. 559–592, 2000.

[Natt 86]   F. Natterer and F. Natterer. *The mathematics of computerized tomogra-
            phy*. Springer, 1986.

[Nero 13]   A. Neroladaki, D. Botsikas, S. Boudabbous, C. D. Becker, and X. Mon-
            tet. "Computed tomography of the chest with model-based iterative
            reconstruction using a radiation exposure similar to chest X-ray exam-
            ination: preliminary observations". *European Radiology*, Vol. 23, No. 2,
            pp. 360–366, 2013.

[Noce 99]   J. Nocedal and S. J. Wright. *Numerical Optimization*. Springer-Verlag,
            1999.

[Noel 13] P. B. Noël, B. Renger, M. Fiebich, D. Münzel, A. A. Fingerle, E. J. Rummeny, and M. Dobritz. "Does iterative reconstruction lower CT radiation dose: evaluation of 15,000 examinations". *Plos one*, Vol. 8, No. 11, p. e81141, 2013.

[Noo 02] F. Noo, M. Defrise, R. Clackdoyle, and H. Kudo. "Image reconstruction from fan-beam projections on less than a short scan". *Physics in Medicine and Biology*, Vol. 47, No. 14, p. 2525, 2002.

[Noo 04] F. Noo, R. Clackdoyle, and J. D. Pack. "A two-step Hilbert transform method for 2D image reconstruction". *Physics in Medicine and Biology*, Vol. 49, No. 17, p. 3903, 2004.

[Noo 07] F. Noo, S. Hoppe, F. Dennerlein, G. Lauritsch, and J. Hornegger. "A new scheme for view-dependent data differentiation in fan-beam and cone-beam computed tomography". *Physics in medicine and biology*, Vol. 52, No. 17, p. 5393, 2007.

[Noo 12] F. Noo, K. Schmitt, K. Stierstorfer, and H. Schondube. "Image representation using mollified pixels for iterative reconstruction in x-ray CT". In: *Nuclear Science Symposium and Medical Imaging Conference (NSS/MIC), 2012 IEEE*, pp. 3453–3455, IEEE, 2012.

[Noo 13] F. Noo, A. Wunderlich, D. Heuscher, K. Schmitt, and Z. Yu. "A non-parametric approach for statistical comparison of results from alternative forced choice experiments". In: *SPIE Medical Imaging*, pp. 86730F–86730F, International Society for Optics and Photonics, 2013.

[Norw 13] J. T. Norweck, J. A. Seibert, K. P. Andriole, D. A. Clunie, B. H. Curran, M. J. Flynn, E. Krupinski, R. P. Lieto, D. J. Peck, T. A. Mian, *et al.* "ACR–AAPM–SIIM Technical Standard for Electronic Practice of Medical Imaging". *Journal of digital imaging*, Vol. 26, No. 1, pp. 38–52, 2013.

[OECD 21] OECD. "Computed tomography (CT) exams (indicator). doi: 10.1787/3c994537-en (Accessed on 2021-10-11)". 2021.

[Oppe 11] A. Oppelt. *Imaging systems for medical diagnostics: fundamentals, technical solutions and applications for systems applying ionizing radiation, nuclear magnetic resonance and ultrasound.* John Wiley & Sons, 2011.

[Padd 14] C. Paddock. "No evidence that CT scans, X-rays cause cancer". Medical News Today, 2014.

[Park 82] D. L. Parker. "Optimal short scan convolution reconstruction for fan beam CT". *Medical physics*, Vol. 9, No. 2, pp. 254–257, 1982.

[Pepe 03] M. S. Pepe. *The statistical evaluation of medical tests for classification and prediction.* Oxford University Press, 2003.

[Pete 81] T. Peters. "Algorithms for fast back-and re-projection in computed tomography". *Nuclear Science, IEEE Transactions on*, Vol. 28, No. 4, pp. 3641–3647, 1981.

[Pick 12] P. J. Pickhardt, M. G. Lubner, D. H. Kim, J. Tang, J. A. Ruma, A. M. del Rio, and G.-H. Chen. "Abdominal CT with model-based iterative reconstruction (mbir): Initial results of aprospective trial comparing ultralow-dose with standard-dose imaging". *AJR: American Journal of Roentgenology*, Vol. 199, No. 6, pp. 1266–74, 2012.

[Pope 07]   L. M. Popescu. "Nonparametric ROC and LROC analysis". *Medical physics*, Vol. 34, No. 5, pp. 1556–1564, 2007.

[Rado 17]   J. Radon. "Über die Bestimmung von Funktionen durch ihre Integralwerte längs gewisser Mannigfaltigkeiten". *Berichte Sächsische Akademie der Wissenschaften*, Vol. 69, pp. 262–267, 1917. Reprinted in J. Radon, Gesammelte Abhandlungen, Birkhäuser Verlag, Vienna, 1987.

[Rama 12]   S. Ramani, J. Fessler, *et al.* "A splitting-based iterative algorithm for accelerated statistical X-ray CT reconstruction". *Medical Imaging, IEEE Transactions on*, Vol. 31, No. 3, pp. 677–688, 2012.

[Roob 10]   C. Roobottom, G. Mitchell, and G. Morgan-Hughes. "Radiation-reduction strategies in cardiac computed tomographic angiography". *Clinical radiology*, Vol. 65, No. 11, pp. 859–867, 2010.

[Ross 69]   K. Rossmann. "Point Spread-Function, Line Spread-Function, and Modulation Transfer Function: Tools for the Study of Imaging Systems 1". *Radiology*, Vol. 93, No. 2, pp. 257–272, 1969.

[Sasi 11]   P. Sasieni, J. Shelton, N. Ormiston-Smith, C. Thomson, and P. Silcocks. "What is the lifetime risk of developing cancer?: the effect of adjusting for multiple primaries". *British journal of cancer*, Vol. 105, No. 3, pp. 460–465, 2011.

[Saue 93]   K. Sauer and C. Bouman. "A local update strategy for iterative reconstruction from projections". *Signal Processing, IEEE Transactions on*, Vol. 41, No. 2, pp. 534–548, 1993.

[Schm 12a]   K. Schmitt, F. Noo, J. Hornegger, K. Stierstorfer, and H. Schondube. "Iterative reconstruction using a pyramid-shaped basis function". In: *Nuclear Science Symposium and Medical Imaging Conference (NSS/MIC), 2012 IEEE*, pp. 3456–3460, IEEE, 2012.

[Schm 12b]   K. Schmitt, H. Schondube, K. Stierstorfer, J. Hornegger, and F. Noo. "Analysis of bias induced by various forward projection models in iterative reconstruction". In: *Second International Conference on Image Formation in X-ray Computed Tomography*, pp. 288–292, Citeseer, 2012.

[Schm 13]   K. Schmitt, H. Schöndube, K. Stierstorfer, J. Hornegger, and F. Noo. "Challenges posed by statistical weights and data redundancies in iterative X-ray CT reconstruction". In: *Proceedings of the 12th Fully Three-Dimensional image Reconstruction in Radiology and Nuclear Medicine*, pp. 432–435, Lake Tahoe, CA, USA, 2013.

[Schm 14a]   K. Schmitt, H. Schöndube, K. Stierstorfer, J. Hornegger, and F. Noo. "Task-based comparison of linear forward projection models in iterative CT reconstruction". In: *Proceedings of the third international conference on image formation in X-ray computed tomography*, pp. 56–59, 2014.

[Schm 14b]   K. Schmitt, H. Schöndube, K. Stierstorfer, J. Hornegger, and F. Noo. "FINESSE: a Fast Iterative Non-linear Exact Sub-space SEarch based algorithm for CT imaging". In: *SPIE Medical Imaging*, pp. 90330Q–90330Q, International Society for Optics and Photonics, 2014.

[Schm 17]   K. Schmitt, F. Noo, and H. Schöndube. "Iterative reconstruction of image data in CT". March 21 2017. US Patent 9,600,924.

[Sch 02] P. Schneider and D. H. Eberly. *Geometric tools for computer graphics.* Morgan Kaufmann, 2002.

[Shep 74] L. A. Shepp and B. F. Logan. "The Fourier reconstruction of a head section". *Nuclear Science, IEEE Transactions on*, Vol. 21, No. 3, pp. 21–43, 1974.

[Sidd 85] R. L. Siddon. "Fast calculation of the exact radiological path for a three-dimensional CT array". *Medical physics*, Vol. 12, No. 2, pp. 252–255, 1985.

[Smit 09] R. Smith-Bindman, J. Lipson, R. Marcus, K.-P. Kim, M. Mahesh, R. Gould, A. B. De González, and D. L. Miglioretti. "Radiation dose associated with common computed tomography examinations and the associated lifetime attributable risk of cancer". *Archives of internal medicine*, Vol. 169, No. 22, pp. 2078–2086, 2009.

[Spie 95] P. K. Spiegel. "The first clinical X-ray made in America–100 years". *AJR. American journal of roentgenology*, Vol. 164, No. 1, pp. 241–243, 1995.

[Stie 04] K. Stierstorfer, A. Rauscher, J. Boese, H. Bruder, S. Schaller, and T. Flohr. "Weighted FBP - a simple approximate 3D FBP algorithm for multislice spiral CT with good dose usage for arbitrary pitch". *Physics in medicine and biology*, Vol. 49, No. 11, p. 2209, 2004.

[Sunn 09] J. Sunnegårdh. "Iterative filtered backprojection methods for helical cone-beam CT". 2009.

[Swen 96] R. G. Swensson. "Unified measurement of observer performance in detecting and localizing target objects on images". *Medical physics*, Vol. 23, No. 10, pp. 1709–1725, 1996.

[Thib 00] J.-B. Thibault, K. Sauer, and C. A. Bouman. "Newton-style optimization for emission tomographic estimation". *Journal of Electronic Imaging*, Vol. 9, No. 3, pp. 269–282, 2000.

[Thib 07] J.-B. Thibault, K. D. Sauer, C. A. Bouman, and J. Hsieh. "A three-dimensional statistical approach to improved image quality for multislice helical CT". *Medical physics*, Vol. 34, No. 11, pp. 4526–4544, 2007.

[Toft 96] P. A. Toft and J. A. Sørensen. *The Radon transform-theory and implementation.* PhD thesis, Technical University of DenmarkDanmarks Tekniske Universitet, Department of Informatics and Mathematical Mod-elingInstitut for Informatik og Matematisk Modellering, 1996.

[Ulzh 09] S. Ulzheimer and T. Flohr. "Multislice CT: Current Technology and Future Developments". In: M. F. Reiser, C. Becker, K. Nikolaou, and G. Glazer, Eds., *Multislice CT*, pp. 3–23, Springer Berlin Heidelberg, 2009.

[Vard 12] V. Vardhanabhuti, B. Olubaniyi, R. Loader, R. D. Riordan, M. P. Williams, and C. A. Roobottom. "Image Quality Assessment in Torso Phantom Comparing Effects of Varying Automatic Current Modulation with Filtered Back Projection, Adaptive Statistical, and Model-Based Iterative Reconstruction Techniques in CT". *Journal of Medical Imaging and Radiation Sciences*, Vol. 43, No. 4, pp. 228–238, 2012.

Bibliography

[Vekl87]   E. Veklerov and J. Llacer. "Stopping rule for the MLE algorithm based on statistical hypothesis testing". *Medical Imaging, IEEE Transactions on*, Vol. 6, No. 4, pp. 313–319, 1987.

[Wang04]   Z. Wang, A. C. Bovik, H. R. Sheikh, and E. P. Simoncelli. "Image quality assessment: from error visibility to structural similarity". *Image Processing, IEEE Transactions on*, Vol. 13, No. 4, pp. 600–612, 2004.

[Wang06]   J. Wang, T. Li, H. Lu, and Z. Liang. "Penalized Weighted Least-Squares Approach to Sinogram Noise Reduction and Image Reconstruction for Low-Dose X-Ray Computed Tomography". *IEEE Transactions on Medical Imaging*, Vol. 25, No. 10, pp. 1272–1283, 2006.

[Wund08]   A. Wunderlich and F. Noo. "Image covariance and lesion detectability in direct fan-beam x-ray computed tomography". *Physics in medicine and biology*, Vol. 53, No. 10, p. 2471, 2008.

[Wund09]   A. Wunderlich and F. Noo. "Estimation of Channelized Hotelling observer performance with known class means or known difference of class means". *IEEE Trans. Med. Imag.*, Vol. 28, No. 8, pp. 1198–1207, 2009.

[Wund12a] A. Wunderlich and F. Noo. "A nonparametric procedure for comparing the areas under correlated LROC curves". *Medical Imaging, IEEE Transactions on*, Vol. 31, No. 11, pp. 2050–2061, 2012.

[Wund12b] A. Wunderlich and F. Noo. "On efficient assessment of image-quality metrics based on linear model observers". *Nuclear Science, IEEE Transactions on*, Vol. 59, No. 3, pp. 568–578, 2012.

[Wund13]   A. Wunderlich and F. Noo. "New theoretical results on Channelized Hotelling observer performance estimation with known difference of class means". *IEEE Trans. Nuc. Sci.*, Vol. 60, No. 1, pp. 182–193, 2013.

[Wund14]   A. Wunderlich. "IQmodelo User Manual", 2014.

[Yu11]   Z. Yu, J.-B. Thibault, C. Bouman, K. D. Sauer, J. Hsieh, et al. "Fast model-based X-ray CT reconstruction using spatially nonhomogeneous ICD optimization". *Image Processing, IEEE Transactions on*, Vol. 20, No. 1, pp. 161–175, 2011.

[Yu12]   Z. Yu, F. Noo, F. Dennerlein, A. Wunderlich, G. Lauritsch, and J. Hornegger. "Simulation tools for two-dimensional experiments in x-ray computed tomography using the FORBILD head phantom". *Physics in Medicine and Biology*, Vol. 57, pp. N237–N252, 2012.

[Zhan06]   Y. Zhang, B. T. Pham, and M. P. Eckstein. "The effect of nonlinear human visual system components on performance of a channelized Hotelling observer in structured backgrounds". *Medical Imaging, IEEE Transactions on*, Vol. 25, No. 10, pp. 1348–1362, 2006.

[Zhan07]   Y. Zhang, B. T. Pham, and M. P. Eckstein. "Evaluation of internal noise methods for Hotelling observer models". *Medical physics*, Vol. 34, No. 8, pp. 3312–3322, 2007.

[Zhan14]   Y. Zhang, S. Leng, L. Yu, R. E. Carter, and C. H. McCollough. "Correlation between human and model observer performance for discrimination task in CT". *Physics in medicine and biology*, Vol. 59, No. 13, p. 3389, 2014.

[Zhen00]   J. Zheng, S. S. Saquib, K. Sauer, C. Bouman, *et al.* "Parallelizable Bayesian tomography algorithms with rapid, guaranteed convergence". *Image Processing, IEEE Transactions on,* Vol. 9, No. 10, pp. 1745–1759, 2000.

[Zieg08]   A. Ziegler, T. Nielsen, and M. Grass. "Iterative reconstruction of a region of interest for transmission tomography", *Medical physics,* Vol. 35, No. 4, pp. 1317–1327, 2008.

In der Reihe *Studien zur Mustererkennung*,
herausgegeben von
Prof. Dr.-Ing Heinrich Niemann und Herrn Prof. Dr.-Ing. Elmar Nöth
sind bisher erschienen:

1   Jürgen Haas   Probabilistic Methods in Linguistic Analysis
                         ISBN 978-3-89722-565-7, 2000, 260 S.   40,50 €

2   Manuela Boros   Partielles robustes Parsing spontansprachlicher
                         Dialoge am Beispiel von Zugauskunftdialogen
                         ISBN 978-3-89722-600-5, 2001, 264 S.   40,50 €

3   Stefan Harbeck   Automatische Verfahren zur Sprachdetektion,
                         Landessprachenerkennung und Themendetektion
                         ISBN 978-3-89722-766-8, 2001, 260 S.   40,50 €

4   Julia Fischer   Ein echtzeitfähiges Dialogsystem mit iterativer
                         Ergebnisoptimierung
                         ISBN 978-3-89722-867-2, 2002, 222 S.   40,50 €

5   Ulrike Ahlrichs   Wissensbasierte Szenenexploration auf der Basis
                         erlernter Analysestrategien
                         ISBN 978-3-89722-904-4, 2002, 165 S.   40,50 €

6   Florian Gallwitz   Integrated Stochastic Models for Spontaneous
                         Speech Recognition
                         ISBN 978-3-89722-907-5, 2002, 196 S.   40,50 €

7   Uwe Ohler   Computational Promoter Recognition in
                         Eukaryotic Genomic DNA
                         ISBN 978-3-89722-988-4, 2002, 206 S.   40,50 €

8   Richard Huber   Prosodisch-linguistische Klassifikation
                         von Emotion
                         ISBN 978-3-89722-984-6, 2002, 293 S.   40,50 €

Alle erschienenen Bücher können unter der angegebenen ISBN im Buchhandel oder direkt beim Logos Verlag Berlin (www.logos-verlag.de, Fax: 030 - 42 85 10 92) bestellt werden.